An Introduction to the Philosophy of Science

An Introduction to the Philosophy of Science

FOURTH EDITION

Karel Lambert
Gordon G. Brittan, Jr.

Ridgeview Publishing Company Atascadero, California

Paper text: ISBN 0-924922-10-9
Cloth (Library edition): ISBN 0-924922-60-5

Published in the United States of America
by Ridgeview Publishing Company
Box 686
Atascadero, California 93423

Printed in the United States of America
by Thomson-Shore, Inc.

To *M*

CONTENTS

Note on the Fourth Edition

The fourth edition is a revision of the third edition. Errors have been corrected. Numerous passages have been rewritten. The entire book has been re-set in a slightly larger format.

PREFACE TO THE THIRD EDITION

A third edition calls for some sort of justification, if not also apology. Ours has three parts. The text continues to be used and various friends and colleagues, some of them users, have encouraged us to revise it. There has been a great deal of important work in philosophy of science since the last edition in 1979, much of it by these same friends and colleagues, which we want to make generally available in an introductory way. Finally, we thought we could write a better book, clearer, more concise, more interesting.

The third edition has been almost completely re-written. The chapter on explanation now includes discussion of the causal-statistical and pragmatic accounts of explanation as well as the traditional "covering law" account and it proceeds by considering a large number of exemplary scientific explanations. We have, in fact, tried everywhere to provide more examples. The chapter on confirmation now includes discussion of the Bayesian and "bootstrap" accounts of confirmation as well as of the "positive instance" account and it provides a very brief introduction to probability theory. Some knowledge of probability theory is now a presupposition of work in philosophy of science, just as it has for a long time now been a presupposition of work in many scientific disciplines. The chapter on theories now includes discussion of the semantic account of theories as well as of the historicist and classical accounts and addresses the question of scientific rationality for the first time. Finally, the chapter on the limits of scientific explanation has been revised to accommodate the new accounts of explanation in the second chapter.

We hope that we have managed to include much, although of course not all, of the significant new literature that has appeared since 1979. While the discussion necessarily is brief, it should indicate what is being done and what directions the discipline is taking. A number of philosophy graduate students have told us, in fact, that they've used our text as a kind of handy compendium. It is certainly intended not only as an introductory account but as a record of continuing developments as well.

We have also tried for greater clarity and concision. The clarity has been purchased in part by an organizational scheme that has each chapter discussing three different views of the concept under consideration and the inter-relations among them. Although some students and instructors will no doubt continue to be bothered by our failure to "take a stand" at the end of the chapters, few should fail to see exactly what the central issues are and what sorts of consideration are relevant to their resolution.

Two sorts of criticism of the earlier editions surface regularly. One is that we (occasionally) misstated a view or the criticism of it and (more frequently) failed to mention some crucial point or author. We hope that there are fewer misstatements this time around. Unfortunately, there is bound to be some disagreement about which the crucial points and authors are; it goes without saying that the material we have included reflects our own interests and evaluations.

The other criticism is that the text is "too difficult." As a profession and an academic community we have apparently come a long way from the standard set by Ernest Nagel's *The Structure of Science*. Some people see in this fact evidence of decline; others welcome it as a sign of liberation from the narrowness of traditional positivism and its methods. We don't intend to comment on the controversy here. At the same time, we want to address the "too difficult" charge. Wherever the difficulty has resulted from lack of clarity, or from unnecessary complication in the discussion or an excessive use of technical vocabulary, we have tried to deal with it. But we have tried not to pretend that the issues are simpler than they are. In an introductory text there is necessarily a certain amount of smoothing of rough edges and rounding off. But there are also and inevitably concepts which are not easily grasped the first time through and arguments that need to be studied in some detail. Three points might be stressed. First, a text is introductory insofar as a student can go from it to a more advanced discussion. We think anyone who had gone through ours can enter directly into much of the contemporary literature. Second, our extensive experience has been that students are intrigued by arguments, no matter how complex, so long as both their premises and conclusions are clearly indicated. Our emphasis is often more on the arguments intended to support the various claims made than it is on the claims themselves. Third, confidence and excitement, both founded on a sense of discovery and mastery, come from working through material that has not in any sense been "watered down." We should expect the same levels of rigor and preparation in philosophy of science classes that we do in science classes. The same sense of accomplishment and delight, at least for many students, will result.

This has been a happy collaboration. Perhaps that is another reason why we continue to revise the book. We have also enjoyed the continuing help and encouragement of friends and colleagues. Many of their suggestions have been included although none of them is to be held responsible for our way with them. In this connection, we would especially like to thank Georg Dorn, Clark Glymour, Henry Kyburg, Roger Rosenkrantz, Wesley Salmon, Erhard Scheibe, Paul Teller, Bas van Fraassen, Jules Vuillemin, and John Winnie. It also happens to be the case that almost all of these have made important contributions to the literature in the recent past. For this, too, we're grateful.

Karel Lambert
University of California,
Irvine

Gordon G. Brittan, Jr.
Montana State University

Preface to the Second Edition

The most difficult problem for a textbook author is what he can presume on the part of the reader. It is doubly difficult for us since the philosophy of science draws both on philosophical method and scientific data, if not also on some logic and mathematics. In any case we have tried to presuppose very little. This is part of the meaning of the work "introductory." It also means that whenever technical terminology is used for the first time, it is to be explained clearly, that many examples are to be given, that ample reference should be provided in the footnotes to further literature on the point under discussion.[1]

[1]Thus, as an introduction to physics, the science around which most of the philosophical issues have crystallized traditionally, we suggest Gerald Holtan, *Introduction to Concepts and Theories in Physical Science*, Addison-Wesley (1953), a text that particularly lends itself to philosophical and historical discussion. A more advanced, more contemporary text with the same merits is W. H. Watson, *Understanding Physics Today*, Cambridge University Press (1963). As an introduction to logic, at least an elementary knowledge of which is indispensable in philosophy generally, we suggest K. Lambert and W. Ulrich, *The Nature of Argument*, Macmillan (1979). As for an introduction to philosophy itself, nothing surpasses reading the classic texts and taking an introductory course. On a more advanced level, there is an excellent discussion of the classic texts form the 17th century to the present time specifically from the perspective of the philosophy of science in Gerd Buchdahl, *Metaphysics and the Philosophy of Science*, The MIT Press (1969).

But the word "introductory" does not mean, as far as we are concerned, that issues must be simplified in a misleading way or that standards of rigor and precision must be compromised. We assume of course that readers will be interested in questions about the nature, role and scope of science, although we have also tried to show why these questions are important. We expect that anyone who takes a course in the philosophy of science based on the present text should be able as a result to join in the discussion as it is presently being carried on in professional books and articles. We hope that some appreciation will emerge that at their advancing frontiers there is no sharp distinction to be made between science and philosophy.

This is a completely revised version of a book by the same title we published in 1970. In the interim there have been reviews in the journals, the comments of friends and colleagues, regular use of the text in our own courses, and the criticisms of various referees. We have tried to learn from the experience; philosophy, like science, is a self-corrective enterprise. In addition to correcting the mistakes and polishing the prose, we have almost everywhere amplified our explanations and worked for greater clarity. In many places (for example, the discussions of the "deductive subsumption" theory of explanation and the "paradoxes" of confirmation) the discussion has been rethought and completely rewritten. There is an additional new chapter on the nature of scientific theories and more illustrative detail throughout.

A major criticism made of the first version is that, with only occasional lapses, we never took a "stand" on any of the issues, preferring instead to outline the debate and indicate the alternatives. We take this criticism seriously. But after a great deal of reflection, we have decided to retain our original strategy. The logical empiricist or "classical" position long dominated the philosophy of science in this century. We begin by setting out its views on explanation, confirmation, theories, and the nature of mathematics, and then discuss objections in each case. Most of these objections are of quite recent origin, some of them being our own, and as far as we know they have not been gathered together in an introductory text in a systematic way. Our concern has been to make clear the pattern of argument and counter-argument. To take a "stand" would be to prejudice the discussion prematurely. It would also force us to agree about certain issues when, even after all these years, we still disagree about them. The more appropriate strategy in an introductory textbook, we feel sure, is to indicate alternative positions on the centra questions in as clear and straightforward a manner as possible and to provide extensive bibliographical references, In order to summarize the discussion and to suggest directions in which it might be continued, we have added "Assessments" to the chapters on explanation, confirmation, and theories. The emphasis in these assessments is on the ways in which the various controversies taken up might be resolved or reconciled.

If the question of background presumptions is the most difficult question for a textbook author, the question of selection is next. Inevitably some issues (in our case, for example, the nature of measurement and functional explanation) are left out. We think those that we have included are particularly important and that they tie together in ways which illuminate the general character of science. It was with this last objective in mind, the attempt to place what it is that scientists do in a general perspective, that we have retained (in a substantially altered form) the final chapter draws out, in ways which have invariably interested our students, the larger implications of points made earlier in the book. As such, it also is a kind of summary.

The third and final question for a textbook author is one of order and arrangement. As before, we believe that a book of this kind should above all else try to make clear the structure of arguments in the philosophy of science. But it is not always possible to isolate an argument; there is often some overlapping, cases in which a point raised in one chapter (for example, the connection between confirmation and theoretical context) inevitably comes up again in another. We have tried to keep the order of the argument as clear as possible, without, however, sacrificing a sense of the complexity of the issues or a feel for the "natural dialectic" in the way they have been discussed.

In the original version of this book we included more material on the behavioral sciences than is usually the case. In this version we have added to the material; among other things, there is a great deal more on the methodological motives of behaviorist psychology.

For their criticism and encouragement we have a number of people to thank, too many to list by name. To them and to students who challenged us every step of the way we are grateful.

Karel Lambert
University of California, Irvine

Gordon G. Brittan, Jr.
Montana State University

Chapter one

Introduction

By and large the interest that philosophers have in science takes three forms. The first involves an attempt to determine the adequacy of the scientific picture of the world, to see whether it can be reconciled with the picture presenting itself to the senses and untutored intelligence, and to discover where its limits might lie. For example, progress in physics in the 17th century in part depended on the assumption that the world was, or was very much like, a vast machine and that to understand the behavior of things was to understand them mechanically, i.e., in terms of the motion and interaction of their parts. Philosophers asked whether this assumption could be made without restriction. The French philosopher and mathematician Rene Descartes, in the fifth part of his *Discourse on Method* (1637), answered that it could not; not all human behavior, linguistic behavior in particular, is explainable mechanically. So, in the view of Descartes and like-minded thinkers, the scientific picture of the world did not adequately depict human beings, however accurate it might be in other respects.

Another assumption commonly made by physicists in the 17th century was that the world is ultimately made up of imperceptible particles which have a certain size and shape, but no color, taste, or smell. But again, certain philosophers questioned whether this could really be the case, or whether we could know that it was, since all our knowledge rests on sense experience and since we cannot sense objects having no color, taste, or smell, let alone objects too small to be seen. David Hume pressed this line of argument in *A Treatise of Human Nature* (1739).

A second form that philosophical interest in science takes has to do with the investigation of special concepts internal to the various sciences or with the analysis of individual scientific theories. A well-known example in the history of science is George Berkeley's examination of Newton's theory of motion in *De Motu* (1726). Berkeley urged, for example, that Newton's concept of force was not generally explanatory and he argued that there was no adequate empirical basis for the concepts of absolute space and absolute time. When analyzed these concepts are shown to be problematic; it is the task of a philosopher, or philosophically-inclined scientist, to resolve the problems by clarifying the concepts or by recommending their replacement. To pick a more recent example, philosophers of psychology are now investigating the concept of "information" in order to determine its explanatory power and the precise force of the claim that basically human beings are little more than information-processing machines.

Finally, philosophers are interested in describing scientific activity *per se*, isolating what they take to be its general features and characteristic forms. This third sort of interest involves the analysis of concepts which are not used so much by scientists in their works as they are by philosophers to describe what it is that scientists do.[1] Since scientists, notably, are believed to *explain* the occurrence of broad ranges of phenomena and to *confirm* their *theories*, philosophers of science study the concepts of explanation, confirmation, and theory.

In this book, the emphasis falls on the third form of interest. The next three chapters deal, in fact, with the concepts of explanation, confirmation, and theory respectively. This emphasis is typical of much recent work done in philosophy of science. Along the way, we give attention to the second form of interest, discussing briefly a number of concepts internal to the various sciences. Finally, in chapter five, "The Limits of Scientific Explanation," we turn to the first form of philosophical interest in science mentioned above. At the same time, it should be clear that these various forms of interest cannot be sharply separated. Any attempt

[1]Sociologists and historians of science also try to describe what scientists do or did; to the extent that a reasonably clear distinction can be drawn, what distinguishes the philosopher of science is his or her concentration on the abstract and general elements in scientific activity using a method which involves the analysis of concepts and the attempt not simply to describe but also to evaluate various scientific developments.

to determine the limits of scientific explanation will depend on how the concept of scientific explanation is analyzed, and so on. What is called "philosophy of science" inevitably involves all three areas in varying degrees.

It should also be clear from the examples mentioned so far that philosophy of science is in large part a reflective activity. It begins with concrete scientific achievements and then proceeds to study the implications, concepts, and methods which these achievements embody and contain. This does not mean that the work of philosophers has no impact on further scientific developments. Often philosophical speculation has led to the creation of new theories and methodologies. In fact, during periods of revolutionary ferment in the sciences, a distinction between scientific and philosophical activity is not easy to draw, as in the case, for example, of Einstein's analysis of the concept of simultaneity; at such times, the scientist as much as the philosopher is engaged in conceptual analysis and an evaluation of foundational questions. It is nevertheless true that philosophy is reflective in character. Indeed, it is often urged that the great period of modern philosophy—from Descartes through Immanuel Kant (1724-1804)—consists in an extended reflection on the scientific revolution of the 16th and 17th centuries and in particular on the synthesis of physical theory brought about by Isaac Newton.

Developments in contemporary philosophy of science have been prompted and guided by another set of revolutionary developments in science and mathematics occurring toward the end of the 19th and the beginning of the 20th centuries. Among the more prominent of these are the breakdown and replacement of Newton's physical theory (in different ways and on different levels by Einstein's theory of relativity and by the quantum theory), the development of new foundations for mathematics and the corollary development of mathematical logic, the development of mechanistic biology, and the emergence of the behavioral and social sciences. Though perhaps not as shattering as the intellectual crisis precipitated by the revolutionary developments of the 16th and 17th centuries, when the English poet John Donne wrote "Tis all in pieces, all coherence gone, all just supply, and all relation," nevertheless they have had a profound impact on our culture and have forced philosophers to re-examine the foundations of scientific knowledge, the nature of scientific explanation, and the adequacy of the scientific picture of the world.

The first comprehensive answer to these newly-asked questions is often referred to as "logical empiricism" (sometimes as "logical positivism"). We will call it the "classical" view. Aspects of it were first sketched by the physicist and philosopher Ernst Mach (1838-1916) and then developed in the first half of this century by, among others, Rudolf Carnap, Herbert Feigl, Carl Hempel, Ernest Nagel, and Hans Reichenbach.[2]

In the following chapters, we will examine the classical view in detail. Since much of this examination will be by way of criticism, it is necessary at the outset to make clear both the main lines of the view and the principal motivations of those who advanced it. If the classical view no longer dominates philosophy of science, the sources of its strength and widespread influence should nevertheless be understood and appreciated.

We can begin by looking more closely at the situation in which the first of the logical empiricists, Ernst Mach, found himself during the last years of the 19th century. For the reasons already mentioned, it was a period of rapid change and intellectual crisis. Two considerations were paramount for Mach. One had to do with the development of the atomic theory of matter. Building on the work of John Dalton (1766-1844), physicists and especially chemists in the 19th century tended increasingly to rely on the atomic theory in framing their explanations. But such reliance raised a problem: Since atoms are not observable, it was asked what sort of evidence could support their introduction and how could the empirical significance of the theories into which they entered be appraised. Mach was very reluctant to accept the atomic theory. As he understood it, atoms had to be unobservable in principle (since they lacked the sensible qualities of color, taste, and smell); i.e., no evidence *could* support their introduction. For him, atoms were metaphysical rather then physical in character and thus, though the objects of speculation, had no place in *science*. This line of argument presupposed a more or less sharp distinction between physics and metaphysics, between science and non-science. Much of Mach's philosophical activity was directed to drawing such a sharp distinction. Once it was drawn, he thought, one could excise such metaphysical elements as atoms, forces, absolute space, and absolute time from generally accepted theories and in this way put science on a firm and thoroughly empirical foundation.

The other consideration uppermost in Mach's mind also forced him in the direction of a distinction between science and non-science. This consideration had to do with the cultural role played by science and is more difficult to explain briefly (although it echoes the "idealist" reaction to the first scientific revolution by such philosophers as Descartes and

[2]Nagel's *The Structure of Science*, Hackett Publishing Co., is perhaps the most comprehensive and systematic one-volume exposition of the classical view.

Berkeley alluded to above). Owing largely to the success of Newtonian theories in explaining and predicting celestial phenomena, and the gradual extension of these theories to new domains, electricity for example, many 19th century thinkers maintained that *any* scientific theory had to be both mechanistic and deterministic, as were Newton's, and that scientific theory thus conceived would shortly be in a position to answer every meaningful question that could be asked about the world, human beings, and the place of human beings in the world. But by the last years of the 19th century, serious doubts had been raised about these claims and about the brave new scientific world they suggested. In the first place, efforts to extend mechanical-deterministic theories to physiology and psychology, particularly in the area of human sense perception, bogged down badly. In the second place, universal application of the mechanical-deterministic picture seemed to allow little room for freedom, dignity, or hope, let alone the claims of traditional religion. In the third place, the advances of science did not, despite the promises of its most enthusiastic promoters, seem to be making human life happier or more co-operative. There was, in short, a strong reaction against science; in the eyes of its critics, it had left behind it a string of broken promises and failure to find a place in its picture of things for human beings or to improve their social condition. Mach and many other scientists of his time were concerned to defend science from these attacks and to counter those theologians, metaphysians, and poets who rushed in to fill the "vacuum" that the supposed "bankruptcy" of science had left open. Once again, Mach's strategy was to draw a distinction between science and non-science. He insisted that science properly understood consisted in the formulation and empirical testing of generalizations in terms of which natural phenomena could be both explained and predicted. These generalizations were, in fact, no more than convenient summaries of our sense experiences. It follows that science as such is not committed to mechanical or deterministic theories (however much certain mechanical analogies might suggest further experiments); failure to find them does not, as a result, impugn the scientific enterprise. Nor does science provide us with some sort of ultimate account of the way the world really is, but rather of the course of our sense experience. Questions concerning the ultimate nature of reality are empirically undecidable and therefore non-scientific. Finally, Mach did not think that science issued in sound cures for social ills. Questions of fact are to be distinguished from questions of value, and social questions are often value-laden. In this way, by getting at what he conceived of as the "core" of scientific activity, Mach hoped to make it secure against the attacks of its critics.

The strategy turns on the distinction between science and non-science. How is it to be drawn?[3] Mach thought that *scientific* statements were inter-subjectively testable and that *scientific* concepts could all be analyzed in terms of empirical observations. Although he didn't put it this way, his philosophical successors combined both theses in the "verificationist" theory of meaning. The theory is not easy to state precisely; many philosophers, in fact, have wondered whether it can be stated in a way that is neither false nor trivial. But roughly put, it is the theory that the meaning of concepts and statements is to be understood in terms of the ways in which their application can be verified. Meaning is understood in terms of the method of verification. A sentence is meaningful to the extent that it is possible, at least in principle, to verify it.[4] To the contrary if a statement cannot be verified, if there is no set of observations we might make that could demonstrate its truth or falsity, then it is meaningless.[5] According to the logical empiricists, many traditional philosophical propositions, concerning the ultimate nature of reality for instance, are unverifiable, hence meaningless. As such, they are "non-scientific."

The verification theory of meaning suggests a twin interest in questions of language and in questions of verification or empirical meaningfulness. The label "logical empiricism" is intended to make this interest explicit. As developed by Mach's successors,[6] it is on the one hand a largely linguistic or *formal* account (using the tools of modern logic) of the concepts we employ to describe scientific activity. To provide an analysis of explanation, confirmation, and theory is to indicate logical relations between types of statements. Or, to put it in a slightly different way, the logical empiricists are largely interested in the *structure* of science. Indeed, they tend to conceive of philosophy as the science of

[3]This is often termed the "demarcation problem".

[4]The sentence "There is life on other planets" has not been verified (or falsified). But we know what it would take to verify (or falsify) it.

[5]Slightly more precisely, the logical empiricists recognize two types of meaningful statement, factual statements (which are verifiable) and nonfactual statements (which are true or false solely in virtue of the meanings of the words they contain).

[6]For a historical account of this development, told from the perspective of two of its most distinguished participants (and in the case of Popper eventual critics), see Carnap's autobiography which prefaces *The Philosophy of Rudolf Carnap*, ed., P. A. Schilpp, Open Court Publishing Company (1963) and that of Popper which similarly prefaces *The Philosophy of Karl Popper*, ed., P. A. Schilpp, Open Court Publishing Company (1965). There is a biography of Ernst Mach in English by John Blackmore, *Ernst Mach*, University of California Press (1972) and an excellent account of the intellectual situation at the time by Mary Jo Nye, *Molecular Reality*, American Elsevier (1972).

science, the form or language of science as the object of its investigations. On the other hand, the logical empiricist position strongly emphasizes the *empirical* character of science. Scientific explanations are empirical insofar as they rest on laws or generalizations which have been confirmed and which lead to predictions which can be verified. Scientific theories derive their significance from the ways in which their vocabularies are given an interpretation in terms of sense experience.

The other leading themes of the classical or logical empiricist position follow naturally from this twofold emphasis on the structure of science and on what (in their view) is empirically meaningful. The position assumes from the outset the adequacy and authority of the scientific picture of the world, within its proper domain,[7] and offers an analysis of explanation, confirmation, and theory which allows us to understand the empirical sources of this adequacy and authority. On the classical view, there is but a single model of scientific explanation, which in large part characterizes science *as science*, and scientific progress is seen as the gradual accumulation of more and better information about the world, or at least of our human experience of it. Theories make objective claims since they can be intersubjectively verified or falsified by matching them with the facts and they are either correct or incorrect.

The challenge to this historically significant view has come from at least three different quarters. There are, first, those who have accepted the general framework and underlying assumptions of logical empiricism, but reject one or another of the ways in which it has been elaborated. The causal-statistical model of explanation to be considered in the next chapter, for example, is advanced as an improvement on the traditional logical empiricist account, but it is not intended to undermine the motives of the traditional account. There are, second, those who think that the interesting and important questions in philosophy of science are not simply structural questions, however useful modern logic might be in their investigation. Those who advance the pragmatic account of explanation, as we shall see in the next chapter, argue that the pragmatic or contextual aspects of explanation are more critical than the formal structure of paradigmatic explanations. There are, third, those historically sophisticated philosophers of science who question the logical empiricist understanding of empirical meaningfulness and, a central issue in the fourth chapter, the relation between facts and theories. For these

[7]And not, for example, as the foundation of ethics and politics. The two crucial distinctions on which the position rests are those between factually and linguistically true statements and between facts and values.

philosophers, theories are in an important sense "subjective" and scientific progress is not characterized by gradual accumulation but rather by revolutionary (and not necessarily "progressive") upheaval. Eventually the debate between the logical empiricists and their critics will turn on what are taken to be the aims and purposes of science and of philosophy as well.

Chapter two

EXPLANATION

1. Introduction

"...the distinctive aim of the scientific enterprise," writes a well known philosopher of science, "is to provide systematic and responsibly supported explanations."[1] But (1) what does "responsible support" mean? and (2) what counts as a scientific explanation? This chapter deals with the second question and the next chapter with the first question.

What counts as a scientific explanation is an important and difficult question. It is *important* for two basic reasons. First, as the remark quoted above makes clear, understanding the very goal of science requires understanding what can pass as a scientific explanation. Second, because of the immense prestige of science, scientific explanations have come to be taken as standards of genuine explanations. An important philosophical task, then, is to ascertain, for example, what there is about astronomical explanations in contrast to astrological explanations that commands attention and respect.

The reasons just expressed are neither the only nor perhaps the most

[1]Ernest Nagel, *The Structure of Science*, Hackett Publishing Co., p. 15.

compelling reasons why the question "What is a scientific explanation?" is important. Thus consider the current debate in the U.S. over creationist explanations of the origin of life. The question is whether they should be taught in the scientific curriculum along with evolutionary explanations of the origin of life. Those who say no insist that such explanations are *not* scientific explanations and therefore have no place whatever in the scientific curriculum. Or consider, finally, the enduring question whether there are limits to what can be explained scientifically. For instance, it is often held that since many human actions are voluntary, scientific explanation of much human behavior is not possible because all scientific explanations presuppose behavior to be determined. The cogency of this argument pretty clearly depends in part on what a scientific explanation really is. Indeed it is an argument which will receive detailed attention in the last chapter of this book.

What counts as a scientific explanation is a *difficult* question because the meaning of the word "explanation" is difficult to isolate. For instance, although one may think initially of an explanation as an answer to a why-question, there are many different sorts of why-question and hence a variety of ways in which they may be answered.

Consider the case of drafting of the United States Constitution. The question "Why did the Founding Fathers insist on particular provisions, e.g., election of the President by an electoral college rather than a direct vote of the people?" might be a request for reasons in support of their action—a *justification* or *defense* of it. An appropriate response would be to cite the arguments given in the Federalist Papers written by Hamilton, Madison, and Jay. One of these was that a "small number of persons, selected by their fellow citizens from the general mass, will be most likely to possess the information and discernment" necessary to make such an important choice. But the question might also be a request for an *explanation* of their actions, some indication of their motives and aims. In this case an appropriate response might be to cite the (much-argued) thesis that the Founding Fathers were chiefly concerned to protect their own economic and property interests and thought that one way to do this was to keep the election of the President in the hands of a small, property-owning group of men.[2]

We will not look, then, for a theory that covers every sort of why-question. Rather, we will begin by fixing on one sort of why-question, and dismiss others as relatively unimportant. This aspect of theory formation

[2]The thesis is put forward in Charles Beard's *An Economic Interpretation of the Constitution*.

is not unusual; all theorizing involves some legislation.

Galileo's theory of motion is a case in point. Aristotle's theory, which it supplanted, had taken as its fundamental question: Why does a body move? The question was guided by the assumption that a body is naturally at rest and, therefore, that its movement requires explanation. From Galileo's point of view, this was an unhelpful question and invited physicists to postulate causal mechanisms where none existed. On his theory the assumption that a body is "naturally" at rest is rejected, and, therefore, there is no reason to ask why it is in motion. Aristotle's question might seem obvious, but declaring it illegitimate, as Galileo did, proved to have extremely fruitful consequences—as we now all know.

We shall also restrict our attention primarily to explanations of events or concrete states-of-affairs. Concrete states-of-affairs are situations in the empirical world such as a person's resisting offers of food, a particle penetrating the nucleus of some atom or a soldier's getting leukemia. They are not states-of-affairs such as a certain number being divisible by 2, the set of real numbers not being mappable onto the integers or a certain proposition implying another proposition.

This further restriction is not trivial. Even among why-questions requiring explanations, the restriction is intended to exclude from primary consideration explanations of scientific laws or theories, and mathematical explanations. So questions like "Why does Snell's law hold?", "Why are discrimination learning experiments good evidence for drive reduction theories of learning?", and "Why is there only one null class?" fall outside our primary scope of concern.

2. The Data

Every scientific theory has a subject matter and also data it must accommodate. The theory of classical mechanics, for instance, had the motion of physical bodies as its subject matter, and among the data it had to accommodate was the fact that the orbit of the Earth around the Sun is not perfectly circular. Similarly, theories of scientific explanation have scientific explanations as their subject matter, and among the facts they must accommodate is the fact that certain cases are generally accepted in the scientific community as legitimate instances of scientific explanation.

What, then, are examples of genuine scientific explanations of events? Ten are detailed below; they certainly do not exhaust the enormous variety, nor are they the most profound instances, of scientific explanation. But they are representative and basic.

Example 1: It is observed that a certain human being suffering from extreme fatigue and lengthy food deprivation shows little or no desire to eat when presented with food. In their classic article "An Explanation of Hunger",[3] the American physiologists W.B. Cannon and A.L. Washburn sought to provide an explanation of this fact. Their explanation was that extreme fatigue inhibits the rhythmic contractions of the smooth muscles in the area of the stomach and in the alimentary canal, the contractions alleged to cause eating behavior. The explanation was, however, defective in one respect; it is now known that it is contractions, not in the area of the stomach, but rather in the duodenum that initiate blood chemistry changes which in turn trip off the central mechanisms leading to eating behavior[4].

Example 2: In a certain metallic substance there is only one atom exhibiting radioactive decay in a specific interval of time, say, in a minute. How is this fact to be explained? An appropriate explanation would be to note that the metal is an alloy of two other metals, one radioactive uranium 238, and the other the stable metal lead 206, and that, therefore, the reason the atom decayed in the stipulated time was because it was a U 238 atom.[5]

Example 3: An airbag is observed to have a lesser volume of air in it 3 hours after having been inflated at a higher temperature. To the question, "Why is the volume of air in the airbag now less than it was at inflation?", an appropriate response might be, "Well, the temperature was 90°F when the airbag was inflated and now it is 70°F. But the Boyle-Charles Law tells us that $PV=RT$. So given that the pressure P remained constant, it is easy to see that the volume of air in the airbag must now be less than it was at inflation."

Example 4: During the test trial of an experiment in a T-maze, a rat, deprived of food for 24 hours, is observed to turn right at the choice-point. In the training phase of the experiment, the rat was run an equal number of times to both end-boxes, but found food only in the right end-box. The explanation of this fact in Tolman's cognitive theory of learning would be that prior to the test trial the rat had acquired a demand for

[3]W. B. Cannon and A. L. Washburn, "An Explanation of Hunger," *American Journal of Physiology*, 29 (1912), pp. 441-454.

[4]See S. P. Grossman, *A Textbook of Physiological Psychology*, Wiley & Sons, Inc. (1967), pp. 323-328.

[5]This example is taken from W. Salmon's, *Statistical Explanation and Statistical Relevance*, University of Pittsburgh Press (1971), p. 57.

food, and during the training trials had developed the expectancy that a right turn in the T-maze would lead to food.

Example 5: Two nerve impulses, I_1 and I_2, in close physical proximity in a neuron arrive within 0.3 milliseconds of each other at the synapse of that neuron. Neither has a local potential quite strong enough to fire a certain adjacent dendrite. Nevertheless, the dendrite in question is fired. Why? Because the local potentials of I_1 and I_2 have summated to a degree high enough to evoke a spike potential in the adjacent dendrite, a phenomenon that will occur in the described circumstances provided the arrival time of the distinct nerve impulses does not exceed 0.5 milliseconds.

Example 6: An airplane making a sharp right turn during a landing attempt suddenly seems to hesitate and then plummets to the ground. The weather is excellent, there is no evidence of an explosion or fire prior to the crash, and the engine was observed to be operating properly. The conclusion of the FAA investigators is that the plane stalled. When asked how it happened, the chief investigator explained that the pilot had put the airplane into a 60° bank—far too steep for the conditions—by applying too much right rudder, too much aileron, and by letting his airspeed drop to 80 mph. When asked why the plane stalled in such an attitude, the chief investigator explained that to maintain lift over the wings in a 60° bank in the type of aircraft involved the airspeed has to be above 84 mph. So the airflow over the top of the wings was interrupted. But, under such circumstances, by Bernoulli's law, the pressure on the top of the wings was greatly increased, thus drastically reducing the lift and causing a stall.

Example 7: An unsheltered soldier, who was two kilometers from the hypocenter of a one megaton atomic blast, gets leukemia some years later. When asked why he got leukemia, his attorney replied to a reporter that the blast released radiation that traversed the space between the hypocenter of the blast and his client, was absorbed by his client's cells, and in some as yet unknown way, led to leukemia. He emphasized that the chances of getting leukemia under the conditions in which the soldier found himself are much greater than in the population at large.[6]

Example 8: Two kilograms of copper at 60°C are placed in three

[6]This example is extracted from W. Salmon's, "Why ask "Why?"? An Inquiry Concerning Scientific Explanation" *Proceedings and Addresses of the American Philosophical Association*, 51 (1978), pp. 683-705.

kilograms of water at 20°C. After a while, water and copper reach the same equilibrium temperature, 22.5°C, and then cool down together to the temperature of the surrounding atmosphere. Why is the equilibrium temperature 22.5°C? The explanation is this. Since the specific heats of water and copper are 1 and 0.1, respectively, and since the Conservation of Energy requires that the total amount of heat be neither increased nor diminished, the heat loss of copper, namely, 0.1 x 2 x (60-T) must be the same as the heat gain of water, namely, 1 x 3 x (T-20), where T is the final equilibrium temperature. And this yields directly 22.5°C as the value of T.[7]

Example 9: A female patient of a certain physician contracts general paresis. When asked what accounted for this tragic event, the physician replied, "Because she had untreated, latent syphilis and only untreated latent syphilitics get general paresis."

Example 10: A certain man recovers from a near fatal case of streptococcus. The physician in charge explains to the man's grateful family that the recovery was due to massive doses of ampicillin.

The preceding ten instances of scientific explanation vary in their style of expression, their detail, and their development. Yet they are, prima facie, genuine scientific explanations nevertheless. A theory of scientific explanation seeks to discover some common thread that binds them together and excludes, say, astrological or religious explanations. Having such a data base is also important for the serious assessment of theories of scientific explanation; general agreement on the data is needed to turn aside arguments for (or against) theories that hinge on mere preferences or purely verbal stipulations. For instance, without the prior determination of such a data base, someone could propose a theory of scientific explanation which might be well nigh irrefutable simply by jockeying around the meaning of the word "scientific" to exclude counterexamples. This does not mean that the data cannot be challenged; it only means that if a datum is challenged, it must result from other nonwhimsical, and, perhaps, overwhelming scientific considerations.

Copernicus, for example, partly for reasons of simplicity, challenged the datum that the Sun is observed by a person on the Earth to move around the Earth. That he was successful is reflected in the fact that we now say that though observation tells us the Sun moves round the Earth,

[7]This example is from Bas van Fraassen's, *The Scientific Image*, The Clarendon Press (1980), pp. 107-200.

the Earth *really* revolves around the Sun. Thus, to say a theory must accommodate the data is to say not merely that it explains the data but, in some cases at least, explains the data away.

3. The Classical Theory of Scientific Explanation

The classical theory of scientific explanation has its roots in Aristotle, and has had many distinguished supporters over the centuries. Earlier in this century, its spirit was sponsored by the eminent logician, H. W. B. Joseph,[8] and by the influential philosopher of science, N. R. Campbell.[9] But by far the clearest, most rigorous, and most extensive treatment of the classical theory is due to Carl Hempel, perhaps the most celebrated philosopher of science in the twentieth century.[10] The presentation of the classical theory in this book relies heavily on Hempel's account.

The essence of a scientific explanation, according to the classical theory, is its *form*. To explain an event or state-of-affairs scientifically is to give a correct *argument*, deductive or inductive, for that event. At least one of the premises of the argument must be a scientific law. If the explaining argument is deductive, then explaining an event scientifically amounts to correctly *inferring* the occurrence of that event from certain facts and laws. If the explaining argument is inductive, then explaining an event scientifically amounts merely to *offering* facts and laws that are related with high probability—at least above chance—to the occurrence of the event to be explained. Providing a high probability basis for an event is weaker than inferring an event because the occurrence of that event does not *follow* from the premises in the correct inductive argument.

The argument-plus-law character of scientific explanation is, according to the classical theory, the uniquely identifying *form* of scientific explanations. It is the insistence on laws in the explanation which has led many to speak of the classical theory as "the covering law model" of scientific explanation. As we shall see shortly, this label is a bit misleading.

By way of example, consider how the classical theory would treat the

[8]See *An Introduction to Logic*, Oxford University Press (1906).

[9]See Carl Hempel, *Aspects of Scientific Explanation*, The Free Press (1965), p. 338.

[10]He and his colleague, Paul Oppenheim, initiated modern discussion of the classical theory in their justly famous study, "Studies in the Logic of Explanation", *Philosophy of Science* 15 (1948), pp. 135-175. Later, in 1965, Hempel collected together numerous essays on, and updated his treatment of, the classical theory in his book, *Aspects of Scientific Explanation*.

explanation of (1) a starving person's lack of desire to eat in the presence of great fatigue, and (2) a person's recovering from a severe streptococcal infection. In the former case, the form of the explanation is as follows:

(1) Whenever an animal is subject to extreme fatigue, the rhythmic contractions in the duodenum are blocked.
(2) Whenever the rhythmic contractions in the duodenum are blocked, the blood chemistry changes initiating appetite are not tripped off.
(3) Person *A* (who is near starvation) is extremely fatigued.
(4) So, person *A* has no appetite.

This explanation has the form of a *deductive argument* in which the conclusion (4) is a valid consequence of the premises (1) through (3). In more technical vocabulary, (4) is called the *explanandum*. (1)-(3) are called the *explanans*. (1) and (2) are general laws and (3) is an *antecedent* or *initial condition*. (1) and (2) "cover" the particular state-of-affairs or events in (3) and (4). To explain the event or state-of-affairs in (4), or more accurately to explain the occurrence of that event or state-of-affairs, then, is to infer it from empirical evidence including at least one general law; here explanation is deductive subsumption under laws.

In the case of the survivor of a severe streptococcal infection, the form of the explanation is as follows:

(1) The probability of a person surviving a severe streptococcal infection when administered a massive dose of ampicillin is much greater than .50.
(2) Person *B* had a severe streptococcal infection.
(3) Person *B* was given a massive does of ampicillin.
(4) So, *B* survived a severe streptococcal infection.

This explanation has the form of an *inductive argument* in which the conclusion (4) is supported with high probabililty by the premises (1) through (3). (4) cannot be inferred from (1) through (3) in this case because, given the truth of (1) through (3), it does not follow that (4) would be true.[11] So to explain the occurrence of the event or state-of-affairs in (4), then, is to provide it with high inductive support based on the empirical evidence in (1) through (3) among which there is a law (in the present case, a probabilistic law); here explanation is inductive

[11]For more on why one cannot infer the conclusion of a correct inductive argument from its premises, see *Aspects of Scientific Explanation*, pp. 53-91.

subsumption under laws.

The classical theory of scientific explanation is motivated by a deep seated conviction about what constitutes an acceptable explanation of an event. The conviction is that any "rationally acceptable answer to the question 'Why did event X occur?' must offer information which shows that X was to be expected—if not definitely...then at least with reasonable probability."[12] To say that X was to be expected is to say that the nonoccurrence of X is excluded either necessarily or with a reasonable degree of probability.

Information conforming to this standard for rational explanation is what constitutes scientific understanding and is enlightening in two important ways. First, it helps one see why explaining something scientifically is regarded as the giving of a correct argument, deductive or inductive, involving an appeal to laws. For instance, in the examples given above, the nonoccurrence of *A*'s not eating and the nonoccurrence of *B*'s surviving the streptococcal infection are ruled out—conclusively in the first case, and with high probability in the second case—in virtue of the explanations being conceived as correct arguments. In other words, the condition of adequacy is guaranteed literally by what it means to be a correct argument, deductive or inductive. Second, it helps one to understand why scientific explanations should be taken as standards of rational explanation. Appeals to God's will or to one's Zodiacal Sign provide no information *showing* that the occurrence of certain events, say, John's death or Mary's success in the stock market, respectively, were to be expected and hence no scientific understanding of these events.

The classical theory of scientific explanation has much to recommend it. First, it fits neatly a wide range of examples. Moreover, scientific explanations typically do seem to appeal, implicitly or explicitly, to laws, be they universal (as in the fatigue inhibiting eating example) or statistical (as in the streptococcal infection example). And, indeed, the classical theory suggests a very important reason why scientists search so persistently for laws; they do so in the interests of explanation. Second, it is a fruitful theory in the sense that with very little modification it can be extended to the explanation of the laws themselves. On this theory, to explain a law is to show how it is inferable from more general laws.[13]

[12]Op. cit., pp. 367-368.

[13]See *Aspects of Scientific Explanation*, pp. 343ff. For an especially illuminating account along classical lines of the explanation of Kepler's laws by Newton, see the chapter on explanation in Joseph's *Logic*.

But difficulties loom.

Consider the case of paresis reported in the data base. In that explanation, the explanandum—a certain woman's contracting general paresis—is explained by appealing to the fact that she had untreated latent syphilis. Yet very few persons having untreated latent syphilis ever get general paresis though everyone who gets general paresis has untreated latent syphilis. So the appropriate inductive argument exhibiting the explanation cannot be a correct inductive argument; the premises of that argument would not support the conclusion of that argument with high probability. The classical theory, it would seem, is too narrow; it does not count the paresis example as a scientific explanation.

Earlier we mentioned that a theory sometimes accommodates a datum not by explaining it, but by explaining it away. This is exactly what the classical theory attempts to do in the paresis example. It holds that the paresis example is not a serious threat to the classical conception of scientific explanation because that example is only a partial explanation. A complete explanation of paresis would cite further factors admittedly unknown in current medicine, but certainly presumed to exist. The addition of this new information in the premises of an inductive explanation of paresis would at the very least supplant the low probability inductive argument with a high probability inductive argument.

For instance, medical science could discover some genetic factor which combined with untreated latent syphilis greatly increases the chances of getting general paresis, say, to a probability of .90. Then the paresis example could be put into the form of a correct inductive argument as follows:

(1) The chances of anyone with untreated latent syphilis and genetic factor G getting general paresis is .90.
(2) Mary had genetic factor G and also untreated latent syphilis.
(3) So, Mary contracted general paresis.

So the belief that the classical theory is too narrow is not warranted by the paresis example.

However, there is another example in the data base that is harder to explain away. It is the case involving the decay of a U 238 atom in a specified interval of time, say, a minute. U 238 has a half-life of 4.5×10^9 years. So the probability of a specific atom of U 238 actually decaying in a specified minute of time is extremely low, and hence not to be expected. Nevertheless, if it did occur, its occurrence would be explained by the fact that it was an atom of U 238, a substance possessing a certain atomic structure and thus subject to spontaneous decay. But then the explaining conditions do not support with high probability the occurrence of the event to be explained, and the requisite inductive argument demanded by the classical theory cannot be a correct inductive argument. However, it is impossible, given the prevailing atomic theory in physics, to supplement the explanation of the decaying atom of U 238 with more information such that the low probability inductive argument can be supplanted with a high probability inductive argument. But then the decaying U 238 example cannot be explained away as an incomplete explanation—*even in principle*.

The classical theory of scientific explanation has also been criticized as being too broad, that it counts instances as scientific explanations that do not or should not be so counted. Here is an example. From ancient times the lawful correlation between the behavior of the tides and the position and phases of the moon was well known. Yet, the objection goes, it would not be regarded as a genuine scientific explanation of the tidal variations in Southern California in June that one could infer them from the lawful relation mentioned in the previous sentence and the fact that the moon was in such and such a position and in such and such a phase. Though this case *would* count as an explanation, given the classical conception, it would not really be so regarded, according to the critic, without the *causal* underpinnings of the correlation between the behavior of the tides and movement and phases of the moon being laid bare. And they were, as we all know now, in Isaac Newton's laws of gravitation.

The reaction of classical theorists to this kind of complaint generally has been twofold. On the one hand, there are those who dismiss the complaint as based on a metaphysical relic disposed of long ago by the philosopher, David Hume. Hume argued that when we ask for the "cause" of something or other we are asking (roughly) for nothing *more than* a regularity under which it may be subsumed. For instance, to ask for the cause of the tide's behavior is to ask for some antecedent event, say, the phase and position of the moon, and a regularity, say, that whenever the phase and position of the moon is such and such, the behavior of the tides is so and so, covering the pair of individual events. For this group of classical theorists, then, the exclusion of explanations via the correlation between the behavior of the tides and the phases and position of the moon, on the ground that it is noncausal, is simply unfounded. On the other hand, there are those classical theorists, who though sympathetic to the *intuition* behind the current complaint that the classical theory is too broad, nevertheless say that the notion of "causal law" or "underlying causal process" is so unclear that to junk the classical theory on such grounds amounts to irresponsible mysticism. But this particular pair of defenses of the classical theory has been seriously undercut by recent analyses of cause and effect—as, indeed, we shall soon see.

A final difficulty for the classical theory of scientific explanation, important to an understanding of the motivation for the next theory of scientific explanation to be presented, remains to be discussed. It, too, is a case intended to show that the classical theory is too broad; it satisfies the conditions of the classical theory, but would not be accepted as a scientific explanation.

A Martian, after some time on the earth, observes that Fred Fox, a 32 year old, healthy male has never gotten pregnant. Asking why this is the case, he is told that it is because Fred takes birth control pills and the chances of anyone who takes birth control pills not getting pregnant is very high. Now this is a correct inductive argument, containing a statistical law, but it surely is no explanation of Fred's avoidance of pregnancy. Men do not get pregnant, pills or no pills. The problem here is that the information supplied does not constitute a scientific explanation, despite what the classical theory claims, because that information is *irrelevant* to Fred's incapacity to become pregnant. A good explanation, of course, would dwell rather on the requisite anatomical structure and conditions needed for any organism to become pregnant and the *fact* that Fred Fox lacked these. So the objection is that the classical theory is too broad—permitting irrelevant information to count as a scientific explanation.

4. The Causal-Statistical Theory of Scientific Explanation

Contributors to the development of the causal-statistical theory of scientific explanation generally share the conviction that explaining an event scientifically is neither a case of deducing nor of inducing (providing a high probability argument for) that event from anything, let alone from a combination of facts and laws. Among the noted scholars who have participated in the development of this theory are Michael Scriven and Richard Jeffrey.[14] But the major proponents of the theory are the late and great Hans Reichenbach and his distinguished disciple

[14]Michael Scriven, "Causation as Explanation", *Nous* 9 (1975), pp. 3-16; Richard Jeffrey, "Statistical Explanation vs. Statistical Inference" in *Essays in Honor of Carl G. Hempel*, edited by N. Rescher, Reidel Publishing Co. (1969), pp. 104-113.

Wesley Salmon.[15] The account to be presented here relies heavily on Salmon's statement of the theory.

Causal-statistical theorists find fault not only with the character of the classical theory but also with its foundations. The classical theory, it will be recalled, is partly founded on a belief in what constitutes a rationally acceptable answer to the question "Why so and so?"; the answer must provide information showing why so and so was to be expected. This "condition of adequacy" is firmly rejected by the causal-statistical theorist on the ground that there are rationally acceptable answers to why-question about *unexpected* events. For example, recall again the question "Why did that atom decay in a certain one minute interval?" The answer "Because it was an atom of U 238, etc." though scientifically (and thus rationally) acceptable does not provide information showing why this quite unexpected event was to be expected!

Concerning the argument-like character of scientific explanation emphasized by the classical theorist, the causal-statistical theorist is equally outspoken. His complaint can be put in the form of an analogy. Spending money is not the same as purchasing an object even though it is often the case that when one spends money, one purchases an object. (Repaying a loan, for example, involves, spending money but not purchasing an object.) Similarly, explaining the occurrence of an event scientifically is not the same as deducing or inducing the occurrence of that event even though it is often the case that when one explains the occurrence of an event scientifically, one deduces or induces the occurrence of that event.

Despite these basic disagreements both schools of thought nevertheless think that scientific explanations are distinguishable both by their form and the kind of information they supply. They differ, of course, on what that form is and on the nature of the information required. What, then, is the causal theorist's picture of scientific explanation?

To explain an event scientifically is to present both the set of factors statistically *relevant* to that event and the causal network underlying the statistical regularities involving those factors and that event.

B is said to be *statistically relevant* to *A* if the probability of *A* given *B* is different from the probability of *A*. For instance, 1,000 milligram injections of ampicillin, one each day for seven days, is statistically relevant to recovering from bacterial pneumonia because the probability of recovering from bacterial pneumonia given seven 1,000 milligrams

[15]Hans Reichenbach, *The Philosophy of Space and Time*, Simon and Schuster (1948); and see Wesley Salmon, "Why ask "Why?"?", 51 (1978), pp. 683-705.

injections of ampicillin is different from the probability of recovering from bacterial pneumonia (in the population at large). In contrast, wearing a gray hat is not statistically relevant to recovering from bacterial pneumonia because the requisite probabilities are not different. A *statistical regularity* involving a statistically relevant factor and the event (to be explained) would be, for example, that the probability of recovering from bacterial pneumonia given seven 1,000 milligrams injections of ampicillin is greater than .98.

To illustrate, consider the leukemia case in the data base. On the causal-statistical theory, to explain scientifically why soldier *S* got leukemia is to present (a) the set of statistically relevant factors, for instance,

> (1) Soldier *S* was 2 kilometers from the hypocenter of the atomic explosion;
> (2) The explosion was a one megaton explosion;
> (3) *S* was unsheltered;

(b)the statistical regularities, for instance,

> (4) The probability of a person's acquiring leukemia given that that person was 2 kilometers from the hypocenter of a one megaton atomic explosion, unsheltered, and so on, is less than 1/100 (which is however higher than in the general population)

and (c) the causal processes and interactions underlying the statistical regularities, for instance,

> (5) High radiation, released in the nuclear reactions in the course of the blast, traverses the space between the blast and *S* (a causal process);
> (6) The radiation striking the cells of *S* interacts with them and is absorbed (a causal interaction);
> (7) The absorption of radiation by *S*'s cells initiates a process leading to leukemia (a causal interaction and a causal process).

As is clear from the above example, reference to a causal link between events has given way to talk of causal processes and causal interactions. So the causal-statistical theory is not open to the charge of irresponsible mysticism leveled by some classical theorists against those who make appeal to a mysterious causal *relation*, to a mysterious link between particular events, in scientific explanations. But this evokes the questions: what is a causal process? and what is a causal interaction?

A *causal process* is a continuous spatial and temporal process. The transmission of radiation between the atomic blast and soldier *S* is an example of such a process. It is continuous because there are no gaps in the process. A noncontinuous process, for example, would be the successive changes in illumination in the lights surrounding a marquee producing the appearance of motion.

It is by substituting causal processes for chains of discrete causal sequences that Reichenbach and Salmon avoid the Humeian criticism of causation mentioned in the previous section. Emphasizing a point anticipated much earlier by Alfred Whitehead (among others),[16] Salmon notes that Hume's criticism works only if the items being related by the causal relation are viewed as discrete point-events.[17]

A *causal interaction* is a relatively brief event at which two or more causal processes intersect. A household example is the collision of two billiard balls.

It is important to distinguish between genuine causal processes and imposters. This is done via the method of *mark transmission*. A genuine causal process is capable of transmitting a mark; pseudo (or nongenuine) causal processes are not, even though the process may be continuous. Thus a beam of light traveling from a flashlight to a white wall is a genuine causal process because a piece of red glass stuck in between the flashlight and the wall yields a beam of light which remains red from the glass to the wall; the beam of light transmits a red mark. On the other hand, the beam of light which moves around a white wall as a result of rotating the flashlight is not a genuine causal process because though continuous it will be red only at the point on the wall where the light first hits the wall after the initial introduction of the red glass into the beam; the red mark is not transmitted through later points where the light hits the wall—unless, of course, the red glass is rotated uniformly with the flashlight.

During discussion of the classical theory mention was made that the label "*the* covering law model" was a misleading way of identifying the classical theory. The reason is that the causal-statistical theory is *also* a covering law model; it demands that the event being explained (for instance, *S*'s getting leukemia) be covered by laws and ultimately by causal laws (for instance, those governing the causal processes and interactions outlined above). The difference between the two theories lies in the meaning of "covered". For the classical theorist the event is

[16]Whitehead, A. N., *Process and Reality*, MacMillan (1929), Part II, Section V.
[17]See "Why ask "Why?"?", p. 690.

covered if it is deducible or inducible with the help of laws; for the causal-statistical theorist "covered" has no such restricted meaning, though its exact meaning has never been clearly explained.

Just as the classical theory was motivated in part by a deep conviction about what a rationally acceptable answer to a why-question is, so the causal-statistical theory rests in large measure on a deep conviction that a scientifically acceptable answer to a why-question has to provide *understanding* of the event to be explained in the sense that it must provide information about the causal network underlying that event. Apparently astrological explanations do not qualify as genuine explanations from the point of view of the causal-statistical theory because they do not provide understanding in the sense just outlined. Appeals to "crucial" astrological events are simply not accompanied by descriptions of the intricate network of causal processes and interactions underlying an event such as in the case of soldier *S*'s getting leukemia.

The causal-statistical theory of scientific explanation is a compelling theory in many ways. It, too, fits a wide range of actual examples of scientific explanations and seems to have a clear advantage over the classical theory in two important respects. First, it accounts neatly for the explanation of unexpected events. Secondly, because of its sophisticated treatment of statistically relevant factors, it excludes irrelevant explanations, such as the "explanation" of a man's avoidance of pregnancy via consumption of birth-control pills. Nevertheless there are difficulties with this theory. Some of them are serious enough to provoke concern that it fits no more than an important subset of actual scientific explanations.

The first criticism is that the theory is too narrow. Consider, for instance, the air-bag example in the data base. The explanation of the variation in the volume of the air-bag from one time to another makes appeal to the Boyle-Charles law. But that law is not a causal law because it exhibits only an invariable relationship between the variables of temperature, volume and pressure at one and the same time. Nor is there any appeal to underlying causal processes and interactions. So there are scientific explanations not so counted by the causal-statistical theory.

Causal-statistical theorists adopt the same attitude toward this case that classical theorists adopt toward the paresis explanation; they hold that it is only a partial explanation, that a complete explanation would involve explicit description of the underlying causal network. And, indeed, in this case, the history of science offers splendid vindication. In the famous reduction of thermodynamics to mechanics, the invariable relationships between temperature, volume and pressure exhibited in the

Boyle-Charles law were shown to depend on a causal network of moving, spherical molecules whose physical encounters produce changes in the paths of the molecules. The whole "bloomin, buzzin confusion" is governed by the basic laws of conservation and energy. It is this causal picture which completes the air-bag explanation outlined in the data base.

But this sort of defense is not very plausible when attention is shifted to the example of the decaying U 238 atom. In this case no mention of an underlying causal network of the sort described by the causal-statistical theory is explicitly made; nor indeed is any such network possible—*even in principle*.

The transmission of radiation through space and time from an atomic explosion to the cells of the unfortunate soldier *S* is an example of a causal process. It is, in other words, a continuous spatial and temporal process. Causal networks, in the causal-statistical conception, involve causal processes so understood. But the prevailing explanation in physics of the decaying U 238 atom is via the quantum theory, a theory in which the notion of a causal process as a continuous spatial and temporal process is rejected[18] and so also the larger picture of a causal network including such processes. So there is no causal information to which appeal can be made to explain away the decaying U 238 example as just a partial explanation; the weight of evidence suggests that the causal-statistical theory of scientific explanation is too narrow.

Also, the belief that the causal-statistical theory fits only a subset of cases widely recognized as genuine scientific explanations is supported by an objection of a more general nature. The objection relies on recent reflections concerning a famous thought-experiment utilized by Einstein to support his conviction that the quantum theory is an incomplete picture of the physical world.

The thought-experiment is known as the Einstein-Podolsky-Rosen experiment; *EPR* for short.[19] Consider the following experimental set-up:

[18]See *The Scientific Image*, p. 122.

[19]J. S. Bell "Bertlmann's Socks and The Nature of Reality," *Journal de Physique*, 42 (1981), pp. 41-63.

Figure 1

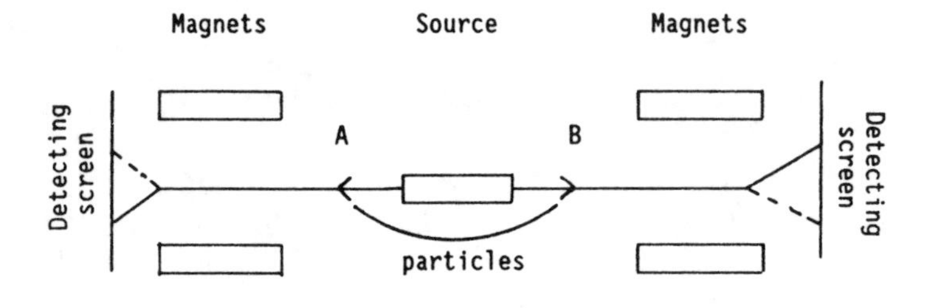

Figure 1 illustrates two suitably prepared particles, *A* and *B*, directed simultaneously from a common source—perhaps as the result of an atomic collision—toward two widely separated magnets followed by detecting screens. Every time the experiment is performed each of the two particles is deflected up or down by the magnetic field on either side of the source. The probability of a particle going up or down on a given occasion is .50 and thus is unpredictable. But when particle *A* goes up on the left, particle *B* goes down on the right, and vice versa; the probability of *B* going down on the right when *A* goes up on the left is 1.0, and the probability of *B* going up on the right when *A* goes down on the left also is 1.0. After a little experience it is enough to look at one side to know all about the other.

The magnets can be varied in numerous ways relative to each other. For example, the magnet on the right can be turned on its horizontal axis 90° while the magnet on the left remains in the same position. Then when particle *A* goes up on the left, particle *B* goes right on the right, and vice versa. So many correlations between the behavior of particle *A* and particle *B* can be established by alterations in the experimental set-up. This experiment has been performed several times,[20] the results conforming in most cases to what the quantum theory predicts.

Einstein believed that the total set of correlations in the *EPR* set-up depended on some "local hidden variable", a "common cause" which enlarged knowledge and more precise description would reveal, and thus believed the *EPR* situation was evidence that the quantum theory was incomplete. But in 1964 John Bell offered a mathematical argument that there can be no local hidden variable to account for all of the possible

[20]Bernard d'Espagnat, "The Quantum Theory and Reality", *Scientific American*, Nov. (1978), pp. 158-181.

correlations in the *EPR* set-up; in short, he presented evidence that there can be no common cause underlying the set of correlations in the *EPR* set-up.[21]

The relevance of the *EPR* experiment and Bell's remarkable proof to the causal-statistical theory is this. Suppose one were asked why particle *B* went down on the right in the unaltered *EPR* set-up. Notice that particle *A*'s going up on the left is a statistically relevant factor because the probability of *B*'s going down on the right when *A* goes up on the left is 1.0 and hence is different from the mere probability of *B*'s going down on the right (which is .50). But on the causal-statistical theory the explanation of *B*'s going down on the right requires elucidation of the causal network underlying the statistically relevant relationship between *A*'s going up on the left and *B*'s going down on the right. Yet Bell's work implies there can be no such causal network. So the existence of a quantum theoretical explanation of *B*'s behavior is evidence again that the causal-statistical theory is too narrow.

A final objection to the causal-statistical theory concerns its goal. It purports to be a theory about what makes for a scientific explanation in general. But it seems on occasion rather to be only an account of "non-superficial" scientific explanations. And, indeed, this appearance is revealed sometimes in Salmon's own language. Thus, in the case of explaining the tides in Southern California by appeal to the phases and positions of the Moon, he suggests that this is a very superficial explanation and that one only obtains significant understanding when one explains in turn the correlation between the tides and the phases and positions of the Moon via Newton's laws of gravitational attraction.[22]

If this is more than oversight on Salmon's part, then the causal-statistical theory would be incomplete because it excludes a certain class of admitted scientific explanations—"superficial" noncausal explanations—from its area of concern. In this respect it would be more deficient than the classical theory.

5. The Pragmatic Theory of Scientific Explanation

The pragmatic theory of scientific explanation was invented by Bas van Fraassen. It is laid out in his book *The Scientific Image*, one of the most influential essays in current philosophy of science.[23] Van

[21] J. S. Bell, op. cit., pp. 41-63.
[22] See "Why ask "Why?"?".
[23] For the reference, see footnote 7.

Fraassen's conception owes much to earlier work on the logic of why-questions, especially to the work of Sylvain Bromberger, C.L. Hamblin, Nuel Belnap, and Bengt Hannson.[24]

The pragmatic theory is at odds with the preceding theories of scientific explanation in three ways. First, the pragmatic theorist is convinced that scientific explanations neither have distinctive form nor supply distinctive information outside of that provided by scientific theories, procedures and facts. To call an explanation scientific "is to say nothing about its form or the sort of information adduced, but only that the explanation draws on science to get this information...".[25] To understand science, then, does not require grasping something over and above its theories, methods and its facts; in particular, it does not require grasping a distinctive *form* of explanation.

Second, the pragmatic theorist is very skeptical about the way in which its competitors apparently distinguish between scientific and nonscientific explanations. Not even all scientific explanations, he notes, have the forms or include the distinctive information claimed by the classical or causal-statistical theories. This point is embodied in the examples adduced earlier to show that both conceptions are too narrow.

Third, the pragmatic theorist rejects outright the view that laws are or should be, invariable ingredients in scientific explanation; in other words, he rejects the covering law picture. The basis of this disagreement with the classical and causal-statistical theories lies in loyalty to the appearances. The pragmatic theorist takes the cases of explanation in the data base at face value, and the data do not always involve explicit appeals to laws. Though not constitutionally against a "deeper" analysis of the data, he is against indiscriminate digging. His goal is not to explain the data away but "to save the appearances."

What then is the pragmatic theory of scientific explanation? It is this: An event or state-of-affairs is explained scientifically if it is the subject of a telling scientific answer to a why-question, where why-questions are identifiable by their topics of concern, contrast classes and explanatory relevance conditions. This looks more formidable than it really is, as an example will make clear.

Consider the question: Why is this conductor warped?[26] *The topic of concern* is that the conductor *is* warped; it is what the questioner supposed to be the case when he asked the question. *The contrast class*

[24]Ibid., chapter 5.

[25]Ibid., p. 155.

[26]The example is van Fraassen's.

of the question is a *set* of alternatives including the topic of concern itself as one member in that set. Thus the contrast evoked by the question could be why *this* conductor here warped in contrast to *that* conductor over there, or why this conductor here warped in contrast to keeping its original shape, and so on. *The explanatory relevance condition* is the respect in which an answer is requested. To illustrate: The request vis a vis the question "Why is this conductor warped?" might be for the events producing the warping of the conductor—for instance, the presence of moisture on the conductor and/or the presence of a magnetic field of a certain strength. Or the request might be for information why the warping of the conductor here as opposed to that one over there *had* to take place—for instance, the information that the warping of *this* conductor (as opposed to that one over there) followed from other facts about the conductor in question and laws relating warping and magnetic fields. Or the request might be for the exact function of the warping of this conductor in, say, the operation of a power unit. And so on...[27] To repeat, a why-question is clarified by specifying its topic of concern, its contrast class, and the respect in which an answer is sought.

The meaning of "telling answer" is best explained by illustration. Suppose we have a Mendelian study on the color of hair which reveals that about 3/4 of the subjects have dark hair, about 1/4 of them have blond hair, and a very small number of subjects turn out to have hair of some other color, say, red. Also, suppose someone to answer the question "Why is this person's hair dark?" as follows: Because she is the offspring of parents from the population in the study in which about 3/4 of them had dark hair and about 1/4 had blond hair. This answer is a telling answer—hence an explanation—to the question because it *favors* the occurrence or state-of-affairs that this person has dark hair—the topic of concern of the question—over the other members of the contrast class, for example, that this person's hair is blond, that this person's hair is red, and so on.

The particular topics of concern, contrast classes and explanatory relevance conditions identifying why-questions are not fixed in advance of their being asked. They can vary from occasion to occasion. The situation is like determining the particular speaker of a sentence beginning with the pronoun "I". One knows, beforehand, that "I" in, say, "I am amazed" refers to the speaker but who the speaker is can vary from occasion to occasion. Determination of who the speaker is depends

[27]Ibid., pp. 141-142.

on the context. So it is with the factors identifying a why-question. It is this parallel that is the source of the label "pragmatic" in the pragmatic theory of scientific explanation.

The pragmatic theory can be illustrated as follows. Consider the example in the data base of the nerve impulse induced in a dendrite. To the question, "Why did that pair of nerve impulses in neuron *N* (in contrast to others arriving at the wall of *N*) evoke an impulse in dendrite *D*?, the scientific explanation is contained in this answer: because that pair of nerve impulses, whose local potentials alone were less than sufficient to evoke an impulse in *D*, arrived at the wall of *N* adjacent to *D* within 0.5 milliseconds of each other and summated to a point exceeding the threshold for evocation of an impulse in *D*. The context of the example made clear (1) what was being contrasted with what—namely, that pair of nerve impulses with other pairs of nerve impulses arriving at the wall of neuron *N*—and (2) the respect in which an answer was requested—namely, for the causes of the induction of an impulse in dendrite *D*. But a slight variation in the context in which the question was asked would show that another kind of answer would be called for (hence another kind of explanation), despite the fact that the contrast class remains the same.

Suppose, instead, the why-question were asked against a background in which the interests of the questioner and respondent had to do with the role that the summating pair of neural impulses played in some muscular response R involving a certain neural pattern containing dendrite *D*. Then the answer requested would not be for the causal connection between the pair of neural impulses arriving at the neuron wall and the evoked impulse in dendrite *D*, but rather for their function in the imagined neural pattern. An adequate answer then might be something like this: to *ensure* that the certain muscular response *R* occurs.

Notice that no appeal to a law is required or explicitly suggested in these answers. Of course, some answers to some why-questions do cite laws—the data base shows that. Recall, for instance the appeal to Bernoulli's law to explain why the airplane stalled. The point is that they are not required on the pragmatic theory because the respect in which the why-question requests an answer may not demand an appeal to laws and/or theories. So, to repeat, the pragmatic theory is not, in any sense of the phrase, a "covering law model".

The pragmatic theory has great appeal for essentially three reasons. First, compared to the other theories of scientific explanation so far examined, it is relatively simple and direct: a scientific explanation is a telling scientific answer to a why-question; no more, no less. Second, it affords a way of accommodating the special aspects of the other theories of scientific explanation; they can be interpreted as addressing special respects in which a scientific answer to a why question is requested. Thus the classical theory can be construed as concerned with the request for information why something had to take place in contrast to something else, and the causal-statistical theory construed as concerned with the request for the causes or underlying causal processes of the occurrence of an event in contrast to the occurrence of another event. Examples of both kinds of request were illustrated earlier in the case of the warping of the conductor. Third, the pragmatic theory has prima facie the broadest scope; it seems best able to accommodate the widest range of cases of scientific explanation without *caveat*. Nevertheless, the pragmatic theory is not complaint free. Some concerns will be detailed below.

The first complaint is that it is unintuitive to say that every why-question asked by a scientist involves a contrast class (other than perhaps, the trivial contrast class reflected in the question: why did event *e* occur in contrast to its *not* occurring?). Consider the case of the Swiss scientist Tremblay who discovered that when you lop off a piece of the organism known as *Hydra*, the piece will completely duplicate the host organism. Imagine now Tremblay asking himself: Why did that piece of *Hydra* duplicate its host? The contrast class couldn't involve other lopped off pieces of *Hydra* because they would duplicate the host, too. And it seems strained to think that Tremblay would have had to have in mind some other organism a lopped off piece of which did not duplicate. And even if he did, he may not have known why the lopped off piece of the other assumed organism did not duplicate.

These considerations at least make one wonder whether the pragmatic theorist has not exaggerated the parallel between assertions such as "I am amazed" and why-questions. For it seems undeniable that when someone normally asserts "I am amazed", there is a referent of the pronoun "I" the contextual determination of which is always essential to understanding the assertion. But the contextual determination of a contrast class for a why-question does not always seem essential to understanding that why-question—as the *Hydra* example is intended to suggest.

To sum up, it is at least disputable that all scientific why-questions require a contrast class. If this is so, then there is a class of scientific explanations—perhaps like the one sought by Tremblay—which do not fit the picture of scientific explanation proposed by the pragmatic theorist.

The second complaint against the pragmatic theory is that it is incomplete. It holds that a scientific explanation is an answer to a why-question, yet there are explanation-seeking questions that do not begin with the word "why" but rather with the word "how". Recall for instance, the example in the data base of the airplane crash resulting from a stall. In that case two questions were asked: Why did the airplane stall? and How did the airplane stall? The answer to the latter is certainly no adequate answer to the former, and vice versa. Intuitively one may be able to explain, say, how an automobile engine works without being able to explain why it works. But if one cannot reduce how-questions to why-questions, then, without amplification, the pragmatic theory is too narrow; it excludes cases of scientific explanation that the scientific community judge to be genuine.

The difference between why-questions and how-questions, indeed, is part of scientific legend. There is, for example, the story of the father of genetics, Gregor Mendel, who, in response to the criticism that his theory of inherited characteristics conflicted with Darwin's views on natural selection replied that there existed no real conflict because he and Darwin were concerned with different questions. Darwin, he said, was concerned with the question why such and such characteristics are inherited, but he (Mendel) was concerned rather with the question how such and such characteristics are inherited.

A third complaint that may be raised against the pragmatic theory concerns the notion of a telling answer. A telling answer, as we have seen, is one that favors the topic of concern of a why-question. But consider the why-question: Why is this person's hair blond? The topic of concern that this person's hair is blond is not favored over the contrast case that this person's hair is dark by the answer "Because she is the offspring of parents from the study population about 3/4 of whom had dark hair and about 1/4 of whom had blond hair." But, as Richard Jeffrey has urged,[28] this answer seems every bit as good an explanation of this person's hair being blond as it is of her hair being dark.

Finally, it may be complained that the pragmatic theory is too broad in the sense that it places no restriction on the respects in which an answer to a why-question can be sought and hence on the sort of answer that can be a scientific explanation. For example, apparently the concern to know the astrological basis of a person's behavior is not ruled out in the question "Why was Fred Fox impulsive?" But surely that concern is

[28]Richard Jeffrey, "Statistical Explanation vs. Statistical Inference" in *Statistical Explanation and Statistical Relevance*, edited by W. Salmon, Univ. of Pittsburgh Press (1971), pp. 19-28.

not a scientific concern—at least in current science.

6. Explanation and Laws

Scientific laws fall into two basic classes; they are *statistical* ("The chances of getting leukemia when exposed to a one megaton atomic blast two kilometers from the hypocenter are less than one in a thousand") or *nonstatistical* ("All copper expands when heated"). Cutting across this basic division is another; laws can be either laws of *coexistence* ("$t=2\pi\ (l/g)$, where t is the period of the pendulum, l its length and g the acceleration of free fall") or laws of *succession* ("Whenever rhythmic contractions in the duodenum occur, blood chemistry changes affecting central mechanisms controlling eating behavior are initiated"). Of noteworthy importance is the fact that laws of succession can be statistical in form ("The probability of a person born in the slums committing a crime sometime in his or/her life is .80").[29] Laws of succession imply a change in the temporal sequence of events; laws of co-existence do not. The relationship between being born in the slums and committing a crime is not a relationship of contemporaneous events; the relationship between the period of a simple pendulum and its length is.

Care must be taken to distinguish between the kind of claim a law makes and the character of the evidence upon which it is based. All laws are supported by evidence that confers at most a high probability on them, but that doesn't mean that all laws are ultimately statistical in form. For instance, the law that all heated copper expands *claims* that any piece of copper when heated expands and not merely that the probability of a piece of copper expanding when heated is practically certain. Indeed, a statement of practical certainty could be true even when a statement of the form "All *A* are *B*" is false. Thus if only one piece of heated copper does not expand the statement "All pieces of copper when heated expand" would be false, but the statement "The probability of a piece of copper expanding when heated is practically certain" would still be true. Clearly the nonstatistical law makes a stronger claim.

Causal laws are a species of laws of succession. They are marked by additional special conditions such as the condition that the events being

[29]The example is Adolph Grunbaum's. See his "Causality and the Science of Human Behavior" in H. Feigl and M. Brodbeck, *Readings in the Philosophy of Science*, Appleton, Century and Crofts (1953), p. 771.

linked in a causal law are spatially contiguous. This "locality" condition is what Einstein felt to be lacking in the EPR set-up discussed earlier; he did not believe in causation at a distance. It occasioned his belief that closer scrutiny of the EPR set-up would reveal the "common cause" initiating the simultaneous but spatially noncontiguous deflections of particles. That is why the suggestion in Bell's argument that there can be no common cause for the total set of EPR correlations has such great philosophical impact.

During the discussion of the classical theory of scientific explanation, the controversial character of causation was mentioned. With the advent of the EPR-Bell story the discussion takes on a new dimension. That story suggests that the very scope of causal laws is limited. An analogy will make the point clear. Snell's law, the law about the angle at which light bends through a medium, was discovered to fail for certain substances, for example, Iceland Spar. Still later Iceland Spar was found to be a nonhomogeneous substance, and Snell's law determined to hold only for homogeneous substances. Similarly for causal laws; the moral of the EPR-Bell story seems to be that causal laws do not hold for very small objects—for example, photons.

This sort of situation has provoked deeper analyses of the EPR-Bell story. Some are investigating assumptions underlying Bell's proof, and others its philosophical implications vis a vis causation.[30] The upshot is that a new aspect has been added to the long controversy surrounding causation and its offspring, causal laws.

It is time now to turn to the pivotal question: What *is* a scientific law? We begin by outlining the context of the question.

Look at the following three statements:

(1) Pieces of copper are physical substances.
(2) Any piece of copper is a good conductor.
(3) All persons sitting on a certain bench in Orange County, California are conservative.

These statements are all of the same form; they are (universal) *generalizations*. That is, they are claims about all members of a certain class—pieces of copper or persons sitting on a certain bench in Orange

[30]See, for example, Erhard Schiebe's essay "What Kind of Hidden Variables are Excluded by Bell's Inequality", *Foundations of Physics A Selection of Papers Contributed to the Physics Section of the 7th International Congress of Logic, Methodology and Philosophy of Science*, edited by P. Weingartner and G. Dorn, Holder-Pichler-Pensky (1986), pp. 251-271, for an example of the former, and Brian Skyrms' "*EPR*: Lessons for Metaphysics", *Midwest Studies in Philosophy*, IX (1984), pp. 245-255, for an example of the latter.

County, California—which are paraphraseable into the form "All *A* are *B*". But only (2) would normally be considered a law of nature. What is the difference? It depends on the pair of statements being considered—the pair (1) and (2) or the pair (2) and (3).

The significant difference between (1) and (2) is that (2) expresses an *empirical* claim but (1) doesn't. The truth of (1) can be established simply from an inspection of the meaning of the word "copper" in a good dictionary. The truth of (2), in contrast, cannot; the empirical facts also play a vital part.

The significant difference between (2) and (3) is that (3)—assuming it to be true—is simply an empirical happenstance, an empirical accident, but (2) is not. Philosophers of science refer to (3) as an *accidental* generalization, but say that (2) is a *lawlike* generalization, that there resides in (2) the trait of *lawlikeness*. So the answer to the question "What *is* a scientific law?" is this: A scientific law is a true lawlike empirical generalization though not necessarily universal in form. The hard problem is to spell out clearly what the word "lawlike" amounts to.

Before turning to the hard problem, let us briefly fix on other aspects of the above definition of a scientific law.

Consider, first, the generalization condition. This means that all laws make claims about a class of things—for example, about pieces of copper or persons sitting on a certain bench in Orange County, California—that are potentially infinite in number. In other words, such statements do not express merely convenient summaries of the known data about particular things—for example, about the expansion of *this* piece of heated coper, *and* the expansion of *that* piece of heated copper, *and* the expansion, ..., copper, etc.

The generalization condition is essential whether the law be nonstatistical (as in case (2) above) or statistical. Thus the statistical law

> (4) The probability of a multicellular organism's getting leukemia when exposed to radiation two kilometers from the hypocenter of a one megaton atomic blast is less than 1/100

is not simply about this or that exposed organism, or about such and such a known set of organisms, but rather is about any exposed organism, even those not yet born.

Next, consider the condition that a law need not be universal in form. This condition is not really new. Laws of statistical form, the reader will recall, are not universal in form; they do not have the form "All *A* are *B*".

Let us now turn to the hard problem, the lawlikeness of some empirical generalizations; (2) is an example of a lawlike generalization, but (3) is not. Note that having the feature of lawlikeness does not mean that the statement is true—or at least not true of the actual world. Ever since the scientific revolution of the 17th century scholars have noticed that the laws of classical mechanics do not describe the phenomena as they actually occur. Accordingly, these statements are not true of the actual world but of an *idealized world* which resembles ours in crucial respects. (This observation raises questions about the sense in which such laws are empirical, but this topic is best left to more advanced discussion.)

What does it mean to say that (2) is lawlike? It means that if an object, which in fact is not a piece of copper, *were* a piece of copper, then it *would* be a good conductor. (3), on the other hand, is not lawlike, so conventional wisdom holds, because it is not the case that if something or someone who in fact is not a person sitting on a certain bench in Orange County, California, *were* such a person, then he *would* be a conservative. A statement of the form "If *A*, which in fact is not the case, were the case, then *B* would be the case" is called a *contrafactual conditional.* So to say a statement is lawlike (that its truth, if it is true, could not have been otherwise) is to say it implies the appropriate contrafactual conditionals.

A concern arises at just this point. The basic complaint is that if lawlike generalizations imply contrafactual conditionals, then it is not clear that lawlikeness is a trait residing in some scientific generalizations but not in others. The operative word here is "residing". To understand the point, a little background is needed.

Ever since Aristotle philosophers and scientists have distinguished between two kinds of properties a physical object can have; properties are *accidental* or *essential.* For instance, human beings have the essential property of being rational, and some have the accidental property of being even-tempered. Again, planets have the essential property of being in motion and some have the accidental property of being molten at the core. For present purposes what is important is the traditional conviction that being accidental and being essential are imprinted on the properties they characterize; they reside or inhere in those properties. With the advent of that peculiarly American philosophy called *Pragmatism*, a different picture of accidental and essential properties, a picture already inherent in Berkeley's views, reemerged. For Berkeley and the pragmatists, all the properties of physical objects are on the same footing; they don't come branded as accidental or as essential. Rather, what is regarded as essential is dependent on communal purposes; essential properties are properties *we* single out as especially important for this or that purpose. Being essential (or accidental) is not an indelible feature of properties.

The skeptic's view of lawlikeness is like the pragmatist's view of essential property; lawlikeness is not a trait imprinted on some scientific generalizations but lacking in others, if lawlikeness is thought of as the capacity to support contrafactual conditionals.

An example will help to make the skeptic's point. Imagine a wall switch containing a copper wire which serves as an electrical conductor when a light switch is thrown. Now consider the lawlike generalization that copper is a good conductor. It implies, so it is argued, that if the light switch (which is now off) were thrown, then the room lights would go on. But imagine that in the room containing the light switch there was the kind of rug that generates a great deal of static electricity when a person walks across it. Then, because of the possible interference from the generated static electricity when a person is walking back and forth in the room, it would not follow, despite the lawlike generalization about copper being a good conductor, that if the light switch were thrown, the room lights would go on. This example shows that contrafactual support is in large part a matter of circumstances. Accordingly, lawlikeness, conceived as contrafactual support, is not a trait that resides or inheres in some generalizations but not in others. So lawlikeness for the skeptic is like being essential for Berkeley and the pragmatists; which generalizations are laws is not dependent on some feature residing in the generalizations so classified. There is an element of decision about which generalizations are laws—or at least that is the pragmatic skeptic's view.[31]

None of this discussion undermines the classical theorist's or the causal-statistical theorist's conception of scientific explanation because it does not challenge the idea that some true scientific generalizations are laws, and others aren't, a distinction which is important to both schools of thought. On the other hand, if lawlikeness is thought to be a trait residing in some generalizations but not in others, then in the absence of some clear characterization of that trait there is an element of mystery—or at least of unmet challenge—in the above theories of scientific explanation.

[31]We owe the current example to Bas van Fraassen.

7. Explanation and Prediction

Until quite recently most philosophers of science espousing the classical theory of scientific explanation believed in a close relationship between explanation and prediction. It was expressed as the doctrine that every explanation is a potential prediction and every (good) prediction is a potential explanation. Both of these claims are now generally regarded as suspect if not outright false. Seeing what led to the change of mind will help to understand how explanation and prediction differ.

A common if somewhat misleading refrain is that one explains the past, but predicts the future. This slogan is misleading because often a future event can be explained by, as well as predicted from, the same information. Consider, for instance, the question: Why will the weather turn bad? On anyone's theory of scientific explanation the following answer is an explanation of that upcoming event: "Because when the atmospheric pressure drops, the weather turns bad, and the atmospheric pressure has dropped." Yet it is clear that from the very same information one can also predict that the weather will turn bad.

Nevertheless the slogan is helpful because it brings out the typical cases of explanation and prediction. Typically one is called on to explain events that have already happened and to predict events that have not yet happened. It is the typical situation, no doubt, which led many proponents of the classical theory of scientific explanation to believe originally that the *form* of an explanation and a prediction are the same, that explanation and prediction differ only with respect to the time at which the event of concern occurs.

Numerous counterexamples to the belief that explanation and prediction are symmetrical have been presented since the belief was first advanced about three decades ago. We present, first, two cases showing that predictions can't always be turned into explanations.

First, recall again the (assumed) true generalization that every one sitting on a certain bench in Orange County, California is conservative. Suppose you observe a person walk up to that remarkable bench and then sit down. It would be reasonable (and justified on the evidence) to predict that that person is conservative. But if asked why that person is conservative, i.e., "What causes that person to be conservative?", it would not be an explanation on anybody's theory to say "Because he is sitting on that certain bench in Orange County, California and all persons sitting on that bench are conservative." The reason why is obvious on the classical and causal-statistical theories both of which are covering law models of explanations; the generalization cited in the alleged explanation is not a law. Not so in the pragmatic theory because it is not a covering law model of explanation, but the answer to the why-question before us now does not constitute an explanation on the pragmatic theory because the context of the question makes only causal answers relevant.

Second, that the information on which a good prediction is based cannot be converted into an explanation is especially clear in certain appeals to statistical data. For instance, on the basis of statistical data, we often predict quite accurately that so-and-so many people will die over high traffic holiday weekends. What is involved is, as in the previous example, a generalization on past data. But intuitively these statistical data do not provide an explanation of, for example, the size of the holiday traffic toll; they do not intuitively yield a relevant answer to the question: Why did so-and-so many people die in traffic accidents over the holiday weekend?

Now consider the belief that an explanation is a potential prediction. Many counterexamples have surfaced since this belief was advanced earlier in this century. And again we will content ourselves here with just three cases.

First, there is the case of Darwin's theory of natural selection. For generations scientists have used this theory to explain the origin, disappearance and variations of species. Yet, apparently, it cannot be used to predict future evolutionary development.[32]

Second, there is the (controversial) case of voluntary actions. Some philosophers believe the explanations of voluntary actions (or more generally of goal-directed or purposive behavior) are never potential predictions. Thus, Hart and Honore claim that "the statement that one person did something because, for example, another threatened him carries no implication or covert assertion that if the circumstances were repeated the same action would follow."[33] In other words, the statement could not be used as a basis on which to predict what will happen.

The argument for this belief about voluntary actions can be put as follows. To explain some voluntary action—the assassination of a president, for example—requires appeal to the assassin's motives, beliefs and/or attitudes—perhaps the assassin's desire for headline fame. But these explanations are not predictive because they are invariably attributed *after* the fact on the basis of the behavior which they are intended to explain. The point has been clearly made by Donald

[32]See Stephen Toulmin, *Foresight and Understanding*, Harper Torchbooks (1963). pp. 24-25.
[33]See their *Causation in the Law*, Oxford University Press (1959), p. 52.

Davidson. He writes:

> ... Generalizations connecting reasons (that is, motives) and action are not, and cannot be sharpened into, the kind of law on the basis of which accurate predictions can reliably be made. If we reflect on the way reasons determine choice, decision, and behavior, it is easy to see why this is so. What emerges, in the *ex post facto* atmosphere of explanation and justification, as the reason frequently was, to the agent at the time of the action, one consideration among many, a reason. Any serious theory for predicting action on the basis of reasons must find a way of evaluating the relative force of various desires and beliefs in the matrix of decision; it cannot take as its starting point the refinement of what is to be expected from a single desire.[34]

Third, and, finally, another difference between prediction and explanation is that the sentence replacing the blank in the answer schema "it will happen that ___" cannot be a law or theory. Though one can explain laws or theories, one cannot predict them. To be sure now and then, one might be able to predict the *discovery* of a law or theory, but the discovery of a law or of a theory is very different from a law or theory; the former, for instance, are events, but the latter are not.

8. Intentional Explanation

An intentional explanation is an explanation containing words like "believes", "desires", "expects", etc. in its explanans. Such words describe processes or acts possessing what the 19th century philosopher, Franz Brentano, called *intentionality*, the property of being aimed at a goal. For instance, to have a desire is to have a desire *for* something and to have a belief is to have a belief *in* something or *that* something is the case.

Example 4 in the data is an intentional explanation. Another example is contained in the answer to the question: Why did the deposed President lie? The answer—because he desired power and he believed he had to lie to get it—is an intentional explanation.

There has been a reluctance by many behavioral scientists to accept intentional explanations. They believe that such explanations commit one to *mental* entities (for example, desires and beliefs) and that claims about such entities are not subject to the kind of inter-subjective verification on which the objectivity of science depends. In the past several decades psychology has been dominated by attempts to remove the taint of mentalism from intentional explanations or to dispense with them entirely. We shall outline the major proposals in a moment, but first it

[34]See his "Actions, Reasons and Causes", *Journal of Philosophy*, LX, 1963, p. 697.

is important to notice that the outcome of the debate has serious implications vis a vis which theory of scientific explanation shall prevail. So we shall first devote some time to this important implication.

Two problems arise for covering law accounts of scientific explanations vis a vis intentional explanations. First, consider the two examples of intentional explanation alluded to two paragraphs back. On the pragmatic theory both examples qualify as legitimate as they stand, but they do not so qualify on the classical theory or the causal-statistical theory because the appeal to a covering law is missing. They are at the most explanation sketches. So an immediate problem is how to turn these explanation sketches into full-fledged explanations. (This problem, and also the next, can be amply illustrated by fixing just on the maze-learning case and the classical treatment of that case.)

The maze-learning case can be turned into a classical explanation in the following way. Assume that rat A is a normal rat.

(1) Rat A demands food.
(2) Rat A expects that if he turns right in the T-maze, he will get the food.
(3) Whenever something that is demanded strongly enough by (normal) organisms is such that some response is expected to result in getting the desired object, then the response will be performed.
(4) Rat A demands food strongly enough.
(5) So, rat A takes the right turn.

This amplified explanation has the form of a classical explanation because (5) is deducible from (1) through (4), and (3) is prima facie a law. So the first problem is easily solved, but that solution gives rise to a second problem, namely, whether (3) really is a law.

Notice that the general form of (3) is: If (normal) organism X has desire Y and belief Z, then X will do or try to do A. One might also want to include a reference to X's perception of his situation as well. The difficulty is not so much that statements of this form generally are false—we all have a great many desires and beliefs on which we do not act—but that they cannot be laws. The reason, as the passage quoted from Davidson a few pages earlier indicates, is that typically we are confronted with a number of conflicting desires. The alleged law in question would have to take such conflict into account. But it is difficult to see how the conflict could be taken into account in a way that did not render the resulting "law" claim trivial. For the resulting claim would have the form: If X has desires P, Q, T and beliefs Y, and if desire P (for action A) is the strongest of his desires, then X will do, or try to do,

A. As a case in point, note in the presumptive law (3) the words "strongly enough". These words serve to pick out the decisive desire alright, but at the expense of trivializing the purported law, for it is trivially true (and hence nonempirical) that when one wants to do something strongly enough, one will do it, or at least try.

To sum up, the important implication of adopting intentional explanations as scientifically legitimate is that covering law models are thereby faced with the vexing problem of finding or amplifying explanations to contain at least one genuine law of nature.

Let us now return to the main question—whether intentional explanations are irrevocably infected with mentalism. The first position is called *reductionistic behaviorism* and in psychology is associated with the great names of Edward Tolman and B.F. Skinner in his earlier writings. They argue that expressions such as "belief", "expectation", "demanded", etc., describe dispositions to behave in certain ways under certain circumstances, and dispositions to behave can be explained in purely behavioral terms, thus removing any allusion, explicit or covert, to the mental.

Take the case of "expectation" for example. The idea is that it is like "solubility" or "magnetic". Soluble sugar dissolves when put in water, and a magnetic iron bar attracts iron fillings when placed near them. Dispositions, in other words, are latent response tendencies. Similarly, an expectant person is a person who responds in a certain way given appropriate stimulating conditions. In particular, consider the following specific example of expectation proposed by Tolman, Ritchie and Kalish in an experimental study done in 1946. They write

> When we assert that a rat expects food at location I, what we assert is that *if* (1) he is deprived of food, (2) he has been trained on path P, (3) he is now put on path P, (4) path P is not blocked and (5) there are other paths which lead away from path P, one of which points directly to location L, *then* he will run down the path which points directly to location L. When we assert that he does *not* expect food at location L, what we assert is that, under the same conditions, he will *not* run down the path which points directly leads to location L.[35]

But serious difficulties confront reductionistic behaviorism. It purports to paraphrase away intentional expressions by means of purely behavioral or physical expressions. Now consider the paraphrase of "expects" mentioned above. There are a variety of circumstances under which,

[35]E. C. Tolman, B. F. Ritchie, D. Kalish, "Studies in Spatial Learning I: Orientation and the Shortcut," *Journal of Experimental Psychology* 36 (1945), p. 15.

despite the conditions mentioned in the paraphrase, the rat will not run down the path—for example, when a cat suddenly crosses it—and yet these are circumstances in which we surely would not conclude that the rat did not expect food at location L.

In fact, the philosopher Roderick Chisholm has urged that every attempt to dispense with intentional expressions must inevitably employ one of them.[36] For example, one might rule out the influence of the cat crossing the path on the rat's behavior by inflating the conditions under which the term "expects" would be truly applicable to the rat. Thus, to the Tolman-Ritchie-Kalish paraphrase one might add that the rat be motivated solely by the demand for food. But "demand" is an intentional expression; and it is Chisholm's view that any attempt to explain away failures of a given paraphrase for a given intentional expression inevitably will employ intentional expressions, expressions such as "demand". If Chisholm is right, then, of course, the reductionistic behaviorist's way of sustaining intentional explanations has to fail.

Another objection to reductionistic behaviorism is that it eliminates mentalism at the expense of causation. The objection, vigorously sponsored by materialists—a position to be examined next—is that dispositions are not causes; solubility, for example, doesn't cause sugar to dissolve. What causes the sugar to dissolve is its being an object of a certain chemical structure placed in water.

Materialism also purports to rid intentional explanation of mentalistic associations. Represented in psychology by many psychologists who claim to be behaviorists, this theory does not construe intentional entities as dispositions to behave in certain ways under certain circumstances, but rather as genuine causes (but not, of course, mental causes). Materialists take expressions like "having an expectation" as simply another way of describing a physical cause. In other words, having an expectation, having a belief, or having a desire are each of them identified with physiological states of one sort or another. When a materialist asserts that having an expectation is identical with a certain physiological state, perhaps at the moment unknown, the identity in question is a factual one, like the identity expressed in the statement, "Tully is identical with Cicero", and not a nonfactual identity like the identity expressed in the statement "Zero is identical with the set which contains the null set as its only member". So for the materialist, the expectation cited in the explanation of the rat's behavior in the maze is at bottom a certain, perhaps still vaguely understood, physiological process.

57

[36]. R. Chisholm, *Perceiving*, Cornell University Press, Ithaca (1975), Chapter 10.

Materialism avoids the earlier objections brought against reductionistic behaviorism because it does not hold that the entire meaning of an intentional expression is captured by a paraphrase in purely behavioral or physical terms. So for the materialist words like "belief" and "demand" form an indispensable part of the theoretical vocabulary used to account for the behavior of at least the higher organisms. But the materialist's conception of the intentional as merely a different way of talking about what is essentially physical has serious problems of its own; it appears to many to conflict with certain ultimate truths.

A basic principle of logic, called the non-identity of discernibles, says that if a given property is possessed by an item *a* but not by item *b*, then *a* is not the same as *b*. The major complaint against materialism utilizes this logical principle. For on the one hand, beliefs have the properties of being true or false, of being justified or being unjustified, of being convincing or being unconvincing, etc. Yet, on the other hand, physiological states are neither convincing nor unconvincing, neither justified nor unjustified, and neither true nor false; indeed, it certainly stretches one's credulity, if not one's sense of coherence, to ascribe, for example, the properties of being true or of being false to physiological states. To be sure physiological states do share many properties in common with beliefs—for example, one can speak of enduring beliefs as well as of enduring physiological states—but the point is that there are many properties beliefs have that physiological states do not have. So, in virtue of the logical principle that discernible things are not identical, beliefs cannot be physiological states of any kind.

The final and most radical attitude toward the place of the intentional in behavioral science is called *replacement behaviorism*. It is most clearly represented in B.F. Skinner's later work. A replacement behaviorist believes it to be impossible to escape the positing of mental states if intentional terms are included in the scientific vocabulary. He frowns on attempts to analyze expressions like "expects" and "desire" in nonintentional terms or to interpret "having an expectation" or "having a desire" as representing physiological states. Rather, one begins with certain notions as primitive—perhaps not even now widely used in the analysis of human behavior—and stakes one's case on the eventual success in controlling and predicting behavior. The whole idea behind replacement behaviorism is to get along without intentional explanations.

The replacement behaviorist holds that for any behavioral occurrence explained using expressions like "expects" and "demands", there is an appropriate corresponding explanation not using such words. Thus reconsider the intentional explanation of the rat's behavior in the T-maze. A behavioral replacement might go thus: Why did the rat turn right in the T-maze? Because the rat has been conditioned to turn right in the T-maze. In this explanation, the technical word "conditioned" occurs, and that expression, on the face of it, is nonintentional.

Replacement behaviorism raises the issue squarely: Can the intentional be avoided in the science of behavior? The replacement behaviorist, of course, answers yes, but the matter is complicated and not easy to decide. For example, there are good reasons to believe that a vocabulary adequate at least to the description of behavior cannot avoid the use of intentional expressions. The unavoidability of intentional expressions in the description of behavior is justified by examples like this. Suppose a person is trained to make a certain response, say, pressing a telegraph key, to a stimulus designated as a line of a certain length. So far the stimulus is characterized completely in terms of nonintentional expressions such as "length". But suppose we embed the same line in a version of the Muller-Lyer illusion and then ask the subject to respond only when he sees a line of the same length as the training line. The subject of course will respond not to a line of the same actual physical length, but rather to a line which looks as long as the original training line. For the nature of this version of the Muller-Lyer is just that two lines of different actual lengths nevertheless can appear to be the same length to the subject. This means, so it is claimed, that the stimulus evoking the pressing of the telegraph key can be characterized adequately only with the help of intentional expressions such as "look as if". And since the description of behavior requires description not only of the actions performed by the organism but also of the environmental circumstances in which those actions are performed, it seems that in this example we have a clear case of the unavoidability of intentional expressions in the description of behavior.

To sum up, two different attitudes towards the place of intentional explanations in the science of behavior have been described. First, there are those, like the reductionistic behaviorist and the materialist, who think that such explanations have an essential theoretical role in behavioral science and seek to legitimize them by eliminating any threat of mentalism. Second, there are those, like the replacement behaviorist, who think that the mentalism latent in intentional explanations is not avoidable. Both attitudes are open to serious criticism, as we have tried to make clear.

9. Explanation and Understanding

Why do scientists acting in their capacity as scientists seek explanations? This is an important question because the answer promises to explain an activity that many scientists and philosophers believe to be the very goal of science—the search for well-founded explanations. Indeed this chapter began with a remark to that effect by one of the most distinguished philosophers of science of our time, Ernest Nagel.

The natural answer to the question—to achieve *understanding*—is not transparent, despite its deceptively simple appearance. The reason is that the notion of *scientific understanding* has less often been the object of methodical study by philosophers of science than the object of (sometimes impassioned) sloganeering. There are notable exceptions—Michael Friedman, for instance.[37] Accordingly we shall begin the penultimate section of this chapter with a discussion of what scientific understanding is not and we shall end with some reasoned speculations about what it is—at least as it bears on the matter of scientific explanation.

First, scientific understanding is not a subjective experience; in particular, it is not a feeling of familiarity nor a sense of relief from perplexity, though the latter experience may often be the result of genuine scientific understanding. These are points that Hempel has made with great eloquence.[38]

Indeed much of what is familiar is in need of understanding. For instance, nothing could be more familiar than the physiological devastation produced by various kinds of cancer. But the world anxiously awaits a good understanding of these phenomena. Moreover, if scientific understanding were simply reduction to the familiar, many explanations by means of relativity theory would have had to be rejected by an earlier generation in favor of explanations by means of the much more familiar Newtonian theory.

There is a parallel between the notion of scientific understanding and the notion of proof. Though it is people who find proofs but differ in their capacities to do so, the word "proof" stands for a perfectly objective kind of thing. Similarly though it is people who acquire scientific understanding but differ in their capacities to achieve it, the expression "scientific understanding" stands for something quite objective. So much is reflected in the fact that when a scientist makes a great explanatory

[37]Michael Friedman, "Explanation and Scientific Understanding", *The Journal of Philosophy*, 71 (1974), pp. 5-19.

[38]See *Philosophy of Natural Science*, Prentice-Hall (1966), p. 83.

breakthrough, one often hears not only that he or she has achieved scientific understanding of some phenomenon but also that *the world* has obtained scientific understanding of the phenomenon in question.

Second, sometimes scientific understanding is thought to consist simply in the accumulation of truths about the universe. Thus one often hears it said that new information about the solar system has increased dramatically because of the Voyager explorations, and so, therefore, has our understanding of that system. To be sure this *is* a conventional sense of "scientific understanding", but it is *not* the sense relevant to scientific explanation; it is not the sense involved when one says scientific explanations yield understanding.

Imagine, for instance, a being of supernatural intellect like that conceived by the great French scientist, Laplace. Imagine this being to know all the laws of nature and all the empirical facts obtaining at a particular time. Now suppose a new fact to be discovered. It is easy to imagine our supernatural intellect nevertheless being in a quandry about how to explain the novel fact. The moral is that though achieving scientific understanding involves the acquisition of new information, not all new information is scientific understanding (in the sense relevant to explanation).

Finally, there is the conception of scientific understanding as the unification of diverse phenomena under a few generalizations or laws. For instance, the molecular-kinetic theory of gases, alluded to earlier in this chapter, provides understanding in the sense that it unifies the behavior of gases under one theoretical roof by connecting such behavior to the behavior of other bodies—molecules—that obey the Newtonian laws of motion. This kind of scientific understanding, carefully spelled out by Michael Friedman, is not normally, however, the sort of information yielded by scientific explanations. The reason is that scientific understanding so conceived is instead a product of *theories*. Though theories often, if not always, provide the bases for scientific explanations, they are not the same thing; one can know the theory of a given subject matter but still not know how it can be used to explain some state-of-affairs.

What then is the new information yielded by a scientific explanation that constitutes scientific understanding? For that matter, what does it mean to say that a scientific explanation yields scientific understanding? There is a uniform answer to the latter question and as many answers to the former question as there are theories of scientific explanation. Let us, therefore, consider the latter question first.

To say that a scientific explanation yields scientific understanding is to say that it *shows* or *exhibits* some new piece of information. This way of talking is common among those philosophers of science who talk seriously about scientific understanding. The important question, then, concerns what the new information shown or exhibited by a scientific explanation that constitutes scientific understanding is.

To answer this question, consider both the question: Why did it rain in Salzburg, Austria on Sunday? and the answer: Because it was 20°C, a front preceding a low pressure area crossed the region containing Salzburg early on Sunday morning, and whenever that sort of thing happens, rain will occur if the temperature is significantly above freezing. Each of the three theories of scientific explanation discussed earlier consider this answer a scientific explanation of the concrete state-of-affairs constituting the topic of concern of the question, that is, that it rained in Salzburg on Sunday. They might all even agree that this explanation *shows* (1) why that topic of concern was to be expected, (2) how that topic of concern was brought about, and (3) why that topic of concern was to be favored in contrast, say, to the states-of-affairs that it rained in Paris on Sunday, that it rained in Vienna on Sunday, and so on. But the three theories disagree over which of the exhibited pieces of information constitutes scientific understanding and hence why the answer in question is a scientific explanation. For a general desideratum in the determination of whether a given answer to a why-question is a scientific explanation is that it yield scientific understanding. So different conceptions about what constitutes scientific understanding are bound to produce different conceptions of scientific explanation. At any rate, the *classical theory*, as we saw earlier, takes (1) to be the kind of information that is scientific understanding; learning why the state-of-affairs that it rained in Salzburg on Sunday is to be expected is the important new piece of information one obtains from the answer to the why-question about that state-of-affairs. For the *causal-statistical theory*, as we saw earlier, it is (2) that constitutes scientific understanding, and for the *pragmatic theory* (3), apparently, is the relevant piece of information constituting scientific understanding.[39]

There is, however, an important difference between the three theories on the relation of scientific explanation to scientific understanding. Notice, first, that in each case it seems that in the preferred view of scientific understanding the following standard holds:

Something is a scientific explanation if and only if it yields scientific

[39]This is admittedly very speculative vis a vis the pragmatic theory because the view of scientific understanding in that theory has never been clearly explicated.

understanding.

In virtue of this standard what and whether something is a scientific explanation depends exactly on one's conception of scientific understanding. Hence detailed accounts of the latter, which in general do not exist, should serve to explain one's propensity to adopt the particular theory of scientific explanation one does adopt.

On the other hand, it is also important to notice that an answer to a why-question may yield scientific understanding, in some recognized sense of scientific understanding, yet not itself constitute a scientific explanation. Van Fraassen, for example, has urged that an answer to a why-question may yield understanding in the causal-statistical sense of showing how a given state-of-affairs has been brought about but still may not constitute a scientific explanation.[40]

To illustrate, consider the pair of questions discussed earlier in the section on the pragmatic theory of scientific explanation, the questions, "Why is this person's hair dark?" and "Why is this person's hair blond?". Van Fraassen believes the answer, "Because she is the offspring of parents from the study population about 3/4 of whom had dark hair and about 1/4 of whom had blond hair" is an explanation only of the topic of concern in the first question but it, nevertheless, yields understanding in the sense of showing how the topics of concern in *both* questions came about. If he is right, then, an answer to a why-question can yield understanding in a sense at least relevant to scientific explanation even though that answer is not an explanation.

The topic of scientific understanding and its relation to scientific explanation is so important and far reaching that it bears a kind of detailed scrutiny long overdue in the philosophy of science. Probably that task has not been undertaken because of the unfortunate tendency to think that talk about understanding is essentially talk about explanation in different words. But if scientific understanding is presented as a *motive* for seeking explanations, this assimilation of scientific understanding to scientific explanation is quite unwarranted.

10. A Final Word

To summarize, consider the question, "Why did Murphy recover from his severe streptococcal infection?", and the answer "Because he was given massive doses of ampicillin and the probability of anyone recover-

[40]See his "Salmon on Explanation", *The Journal of Philosophy*, 11 (1985), pp. 639-651.

ing from a severe streptococcal infection when given massive doses of ampicillin is very high." The three theories of scientific explanation examined in this chapter all count the above answer to the given why-question a scientific explanation but differ on the features of that answer which make it so. The classical theory contends that the answer qualifies as a scientific explanation because the state-of-affairs expressed in the why-question is, loosely speaking, inferable from the answer. The causal-statistical theory holds instead that it is because the answer delineates statistically relevant factors vis-a-vis recovery from a severe streptococcal infection—for instance, receiving massive doses of ampicillin—and, if somewhat vaguely, the underlying causal network that it counts as a scientific explanation. Finally, the pragmatic theory asserts that the answer qualifies as a scientific explanation because it favors recovery from a severe streptococcal infection over alternative states-of-affairs—for instance, the recovery of O'Brien, who received no ampicillin, from a severe streptococcal infection.

The differences between the various theories have been sketched in the earlier pages of this chapter as well as some of the commonly expressed strengths and weaknesses of each. Moreover, there has been a discussion of certain implications of these theories—some general, some specific—for topics such as natural law, prediction, intentional explanation and scientific understanding. These theories all raise questions; any good theory does. Their worth and fate depend on how their supporters respond to these objections, and they all have very able supporters. For the interested reader, their (often ingenious) responses are available in other places.[41]

Moreover the concern in this chapter has been solely with the question, "*What* is a scientific explanation?". It has not concerned itself with the vitally important questions, "What is a *correct* explanation?" and, "What is a *good* explanation?", two by no means equivalent questions which are subtopics in the area of concern called the *evaluation* of scientific explanations. For the interested reader these topics are treated with clarity and ingenuity by Peter Achinstein in his book *The Nature of Scientific Explanation*.[42]

[41]See, especially, C. G. Hempel, "Postscript" in *Aspects of Scientific Explanation*, De Gruyter (1977), pp. 97-123; W. Salmon, *Scientific Explanation and the Causal Structure of the World*, Princeton Univ. Press (1984); and B. van Fraassen, op. cit.
[42]Oxford University Press (1983).

Chapter three

CONFIRMATION

1. Introduction

Scientists advance hypotheses and carry out experiments. These activities are closely connected. Often hypotheses that, in one way or another, help to initiate experiments are accepted provided they in turn are confirmed by the results of those experiments.

The central problem arises in this way. Many of the claims we make "go beyond" the current evidence for them in the sense that the evidence may be perfectly good, yet the claim nevertheless be false. Human beings uniformly experience the daily rising of the sun and yet the claim that the sun will rise tomorrow could very well be false. Correct arguments whose conclusions must be true if their premises are true are—as noted in the chapter on explanation—*deductively correct*. Correct arguments whose conclusions are possibly false when their premises are true are *inductive*; their premises *support*, but do not *entail*, their conclusions. The sample argument above whose conclusion is that the sun will rise tomorrow is inductive; the premise only supports the conclusion.

For the moment we will restrict our attention to scientific hypotheses that have the form of universal generalizations, the form "All A are B." Of course many scientific hypotheses do not have such a simple form, but in what follows nothing turns on the point. Hypotheses of the form

"All A are B" inevitably "go beyond" the evidence for them. Up until the voyages of the celebrated Captain Cook (1728-1779), all the swans that had been observed were white. Nevertheless the universal generalization "All swans are white" was false since Cook found black swans in Australia. At any given time one's observable evidence is never more than partial and incomplete; yet hypotheses of the type indicated are completely general. So arguments intended to establish universal hypotheses on the basis of such evidence must be inductive. The premises stating the evidence can do no more than support the conclusion. Naturally interest centers on those hypotheses that are well supported rather than poorly supported. We will say that those hypotheses that are well supported are well *confirmed*. The problem of confirmation is the problem: When and under what conditions does evidence confirm a hypothesis?

The "problem of confirmation," is *not* the famous philosophical "problem of induction." The problem of induction was first clearly formulated by David Hume (1711-1776) in *A Treatise of Human Nature*. It has to do with the *justification* of inductive reasoning. Why, Hume asked, it is ever rational to accept the conclusion of an inductive argument? One can't say that the conclusion *follows* from the premises, for in that case the argument would be deductive and not inductive. On the other hand, one can't say that one's past experience supports accepting the conclusions of inductive arguments, for that would be to reason inductively and, therefore, in a circle. But there is nothing else to say by way of justification. So, Hume concluded, it is never rational to accept the conclusion of an inductive argument. If one accepts them nonetheless (we all believe, after all, that the sun will rise tomorrow), it must be on other sorts of (non-rational) grounds. Since Hume thought that most if not all of our empirical knowledge depends on inductive reasoning, and since he thought he had shown that such reasoning could not be justified, he counseled skepticism; even those scientific claims about which we are most convinced are no more than a matter of habit and belief.

Hume's argument has been extremely important in the history of philosophy. There is no easy answer to it. But it can be by-passed here. In the first place, one can distinguish between those hypotheses which have been *accepted* in the history of science—let's say since the 17th century—and those which have not. Our question is whether those which have been accepted bear some common relation to the evidence for them. In this context, the problem of confirmation is to determine (whatever the process) whether there is such a relation and, if so, to make its structure clear. In the second place, one can distinguish between those hypotheses which, intuitively, are supported by the evidence and those which are not. In this context, the problem of confirmation is to analyze the intuitions involved and to see whether they can be refined and systematized into a general account of the relation between evidence and hypothesis. The point is that working with actual practice and self-conscious intuition one may be able to reach some sort of reflective equilibrium. If it is never *rational*, as Hume insists, to accept a general hypothesis, scientific or otherwise, it is still possible to distinguish between those which are better and those which are worse confirmed. Our interest, of course, is in those which are better confirmed. Vis-a-vis the eventual criterion of rationality, this enterprise is important because it depends inevitably on factors internal to science and reflections on them.

2. Confirmation versus Corroboration

Some philosophers, notably Sir Karl Popper in his very influential work *The Logic of Scientific Discovery* (1935; English translation, 1959) have urged that the emphasis on the confirmation of hypotheses is badly misplaced. Rather, one should concentrate on their falsification. Popper's view merits some consideration. It rests on three basic points.

The first point is purely logical. There is an important difference between confirming and falsifying hypotheses that are universal generalizations. No number of confirming instances, no matter how great, can show that a universal generalization is true. Confirmation is inevitably indecisive. Yet a single disconfirming instance, other things being equal, will show, in a deductively correct way, that the generalization is false. It doesn't follow from the fact that a number of cats are curious that all cats are, but it does follow from the fact that there is an uncurious cat that the generalization "All cats are curious" is false.

The second point has to do with actual scientific practice. Scientists do not keep repeating the same experiments in the attempt to pile up confirming instances of a hypothesis. Once an experiment has been performed, and perhaps duplicated, scientists move on to devise new experiments, trying to *test* the hypothesis. But, Popper contends, to test a hypothesis is to try to find ways in which it might be falsified. Since no hypothesis can ever be *established* as true, the best one can say of a particular hypothesis is that it has survived a number of tests, the more varied and severe the better. A hypothesis which has so survived is said to be *corroborated*. To say that a hypothesis H is corroborated is to say that it has not been shown to be false; to say that H is more highly corroborated than H′ is to say that H has survived a greater number of, and/or more severe, tests than H′. Even though H may be a highly corroborated hypothesis, it does not follow that H is true; though, for instance, Snell's initial hypothesis about how the direction of a light ray is affected by a substance through which it passes had survived many tests, and hence was highly corroborated, when tested with the substance Iceland Spar it turned out false. Corroboration is a weaker notion than confirmation because even if a hypothesis H is highly corroborated, it does not follow that H is confirmed; whereas confirmation bears on the truth of a hypothesis, corroboration does not. This is, traditionally, what characterizes the scientific mind: never to accept some truth as given but constantly to question it.

The third point has to do with an intuition underlying a good deal of testing practice. Let us call it the "testing" intuition. It is that hypotheses which have improbable consequences vis-a-vis their rivals are much more highly corroborated if their consequences are not falsified. For example,[1] the English astronomer Edmond Halley first observed a large comet in 1682. Going back through the records he found reports of observations of previous comets precise enough to compare to his own. The orbits recorded for two of these, one in 1606-1607 and the other in 1530-1531, were very close to that which he calculated in 1682. Arguing that it was unlikely that three comets should have such very similar oribts, he concluded that three appearances of a single comet had been observed. Then, using data from what he took to be three appearances of a single comet, together with the hypothesis that Newton's laws applied to the phenomenon, he predicted that the comet would appear again in December, 1758. The comet reappeared as predicted on Christmas Day, 1758, fifteen years after Halley's death, and was promptly named "Halley's Comet." We have just witnessed its return.[2]

Now the point is this. The degree of corroboration of Halley's hypothesis concerning the applicability of Newton's laws to the comet phenomenon depended on the high probability that his prediction was false. Everything known at the time, with the single exception of Newton's Laws, made it highly likely that a comet would not appear within a thirty day period fifty-three years after Halley first made the

[1]Following Ronald Giere, *Understanding Scientific Reasoning*, second edition, Holt, Rinehart & Winston (1984), pp. 97ff.

[2]Although that return did not serve to confirm further the Newtonian hypothesis on the basis of which it was first predicted.

prediction. It was this fact that made the prediction a significant test ("a real test"). Hence, the traditional methodological maxim "Nothing ventured, nothing gained!" So hypotheses with improbable consequences are highly corroborated if their consequences are not falsified. This is, according to Popper, the dynamic of modern science: It consists of bold conjectures such as Halley's and intricate attempts at refutation. Since there is no way to establish a hypothesis, because at best certain hypotheses have so far not been refuted, we can never rest content with what we have.

Any adequate account of confirmation must accomodate the "testing" intuition. But the other two points Popper makes are debatable. In the first place, whatever purely logical advantage a program of falsification enjoys is only temporary. The French physicist, philosopher, and historian of science Pierre Duhem (1861-1916) was perhaps the first to point out that hypotheses are never tested in isolation; their empirical or observational consequences are derived only with the help of other hypotheses, together with so-called initial and boundary conditions. A negative result does not by itself show which of these hypotheses or conditions is to be called into question. Duhem takes as one of his examples the apparent refutation of the particle theory of light.[3] The particle theory consists of a number of hypotheses: that light is formed of particles, that these particles are projected at high velocities by luminous objects including the sun, that these projectiles penetrate all transparent bodies, etc. The various hypotheses taken together imply that "the index of refraction of light passing from one medium into another is equal to the velocity of the light projectile within the medium it penetrates, divided by the velocity of the same projectile in the medium it leaves behind." This proposition in turn implies that light travels faster in water than in air. Foucault (1819-1868) carried out an appropriate, if also somewhat difficult, experiment and found that in fact light travels more slowly in water than in air. But what should we conclude? The negative result might be attributed to the falsity of any one of the hypotheses constituting the particle theory, or to some other assumption about the various media used in the experiment. Schematically, from hypotheses H_1 and H_2 and...and H_n and initial and boundary conditions C_1 and C_2 and...and C_n we derive an experimental consequence E. If E turns out not to be the case, that is, *not-E* is observed, then it follows logically that H_1 is false or H_2 is false...or H_n is false or C_1 is false or..., etc. A negative instance does

[3]*The Aim and Structure of Physical Theory*, Atheneum Publishers (1962), pp. 186ff.

not by itself show *which* one of the hypotheses or initial conditions is false, and hence one has to fall back on other decision methods, confirmation very possibly among them. To put it another way, the difference to which Popper draws attention is matched by another: A confirming instance confirms *all* of the hypotheses and conditions from which it follows; a falsifying instance does not similarly falsify all of the hypotheses and conditions from which *it* follows.[4]

Popper's other point concerns scientific practice. He is surely correct that scientists are interested in testing hypotheses and that they are not interested merely in piling up positive instances. But it is wrong to suggest that tests never have anything to do with truth and hence that hypotheses are never confirmed in any reasonable sense of the word. No interesting hypothesis can ever be established conclusively, once and for all. But it can be confirmed. Thus Halley's test of Newton's theory: he confirmed it. And the more varied the tests it survived, the better it was confirmed. In fact, one can distinguish in actual scientific practice between experiments that confirm and those that falsify, a distinction that on Popper's view should be impossible to make. Thus, observations of the bending of the sun's rays made by Arthur Eddington (1882-1944) during the eclipse expeditions of 1919 confirmed the hypothesis of general relativity.[5] Lavoisier's (1743-1794) experiments on combustion towards the end of the 18th century, on the other hand, falsified the phlogiston hypothesis.[6]

Moreover, there do not seem to be rejection procedures of standard sorts for hypotheses, even when falsifying evidence is at hand. Yet another case from the history of astronomy should help to illustrate the point.[7] In the middle of the 19th century, astronomers were faced with

[4]Popper recognizes the difficulty. It is always "in principle, possible to question the refutation of any hypothesis by asking whether responsibility for the outcome of the refuting experiment might not be attributable to one or more of the cooperating hypotheses," In "Intellectual Autobiography," P.A. Schilpp, ed., *The Philosophy of Karl Popper*, Open Court Publishing Company (1974), p. 1035. Yet Popper does not resolve the difficulty beyond leaving it up to "the scientific instinct of the investigator." See *The Logic of Scientific Discovery*, Hutchinson and Company Limited, p. 76n.

[5]An hypothesis which so far it has been very difficult to test. Its confirmation and eventual acceptance rests in fact on no more than a small number of tests. If one were to insist with Popper that Eddington's observations should be taken as corroborating rather than confirming (since they involved the test of a novel prediction), one would still have to allow that the subsequent experiments constituted confirmations.

[6]This is the hypothesis that various facts about combustion can be explained by postulating the existence of phlogiston–fire stuff–in all combustible materials.

[7]See Martin Grosser, *The Discovery of Neptune*, Harvard University Press (1962).

the fact that irregularities in the motion of the planet Uranus could not be explained using Newton's Laws of Motion and Gravitation in terms of the gravitational attraction of the sun and the other known planets. There were three prominent responses to this situation. One was to say, as Popper would, that Newton's theory had been disconfirmed; alternative views would have to be sought. A second response was to say that the range of the Law of Gravitation, the inverse square law, had to be restricted; the hypothesis satisfactorily explained planetary motion out to a distance from the sun just short of Uranus, and past that point the motion of the planets was describable by another hypothesis—a little mathematical ingenuity sufficed to frame a hypothesis that did account for Uranus' irregularities. The third response, made independently by Adams (1819-1892) in England and Leverrier (1811-77) in France, was to postulate the existence of an undiscovered planet with the specific mass and location, hence with the appropriate gravitational attraction, that would explain the irregularities in the orbit of Uranus. Adams and Leverrier were spectacularly vindicated, as was the Newtonian view they were trying to save, when a short time later the postulated planet, Neptune, was observed through a telescope. Prior to that vindication, the irregularities did not dictate a unique response. There were many different options with respect both to theory and to evidence. In particular, in opposition to what Popper's emphasis on falsification would lead one to expect, rejection of Newton's theory was the least widespread. There was no doubt about the irregularities, nor about the fact that the unamended inverse square law could not deal with them. But in the absence of alternative hypotheses that can explain the phenomena in at least as satisfactory a way as the prevailing theory, scientists hesitate to abandon the prevailing theory. Indeed, even when there are such alternative hypotheses, there is reluctance to abandon particularly well-entrenched theories despite accumulation of apparently falsifying evidence. Instead, efforts are made to accommodate or explain away the evidence.

3. The "Positive Instance" Account of Confirmation

The evidence for an hypothesis can be direct or indirect. *Indirect evidence* is evidence accruing to an hypothesis via its logical connections with other highly confirmed hypotheses or accepted theories. For example, in the psychology of learning, results in the famous latent learning experiments were often taken by cognitive psychologists to provide indirect evidence for the hypothesis that learning is not a gradual accumulation of stimulus-response bond strength, but rather is the discontinuous and often sudden acquisition of a "cognitive map." In this case, the cognitive theorists' hypothesis about the existence of latent learning is believed to be connected theoretically to the non-continuity of learning just described. *Direct evidence*, on the other hand, has been characterized in several different ways. We begin with what was for a long time the standard account.[8]

On this account, developed by Hempel, the direct evidence *E* for a hypothesis *H* consists of its *positive* and *negative instances*. If an object *a* has properties *A* and *B*, or satisfies the description "is A and B," then *a* is a positive instance of the generalization "All A are B." If an object *a* is *A* but not also *B*, or satisfies the description "is A and not B," then *a* is a negative instance of the generalization. If *a* is not *A* to begin with, then *a* is neither a negative nor a positive instance of "All A are B." Suppose that *H* is "All UCI students are wealthy," and suppose that John is a wealthy student at UCI, that Bill is a poor student at UCI, and that Charlie is not even enrolled. Then John constitutes a positive instance of *H*, Bill a negative instance of *H*, and Charlie is neither the one nor the other. The direct evidence *in favor of H* is the set of its positive instances and the direct evidence *against H* is the set of its negative instances. Thus hypotheses can be said to be *confirmed* by their positive instances and *discomfirmed* by their negative instances.

This account has a great deal to recommend it. It is clear, precise, and completely general. Moreover, the basic idea has a long history going back perhaps as far as Aristotle. The idea is often referred to as *induction by simple enumeration*. One supports a generalization by enumerating positive instances of it; the corresponding inductive argument has descriptions of these instances as premises and the generalization as conclusion. The only requirement is that the instances make the generalization more likely, if true, and entail its falsity, if at least one of them is false.

The method is often used to establish causal connections. When two properties are constantly conjoined, one is ready, other things being equal, to infer some sort of causal relationship between them. If a piece of blue litmus paper is dipped into some acid and turns red, then, if the same thing happens on repeated occasions and we discover no other factor in the situation that might have brought about the same result, the conclusion that the acid *causes* the litmus paper to turn red typically is

[8]See Carl Hempel's papers "A Purely Syntactical Definition of Confirmation" and "Studies in the Logic of Confirmation," both reprinted in *Aspects of Scientific Explanation*.

drawn. This is not to suggest, of course, that constant conjunctions or correlations are tantamount to causal connections. In the chapter on explanation we noted some reasons why they cannot be identified. But the evidence for the claim that there is a causal connection between properties is supplied initially by our observing correlations in carefully controlled circumstances.

The underlying intuition is this: Test a generalization by looking at *samples* of it since, by the very nature of the case, it is impossible to *prove* that the generalization is true. Other things being equal, positive samples or instances of a generalization confirm it; the more such samples we have observed, the more highly confirmed it is. This seems to be as much common sense as science.

But however natural or obvious the "positive instance" account of confirmation might seem, there are profound difficulties with it. These difficulties can perhaps best be made clear in connection with two notorious "paradoxes" of confirmation. The first paradox suggests that the "positive instance" account raises questions concerning what counts as a *confirmable hypothesis*. The second paradox leads to problems in the very concept of a *positive instance* itself.

The first paradox, known as the "grue paradox," is due to Nelson Goodman.[9] He develops it as follows. Suppose that all emeralds examined before a certain time *t* (say, the year 2000) are green. Then the observation of green emeralds, if nothing else, confirms the general hypothesis that all emeralds are green. So far, so good. Now suppose a new predicate, "grue," is introduced, a predicate which applies "to all things examined before *t* just in case they are green but to others just in case they are blue." Consider the two hypotheses:

H.1 All emeralds are green.
H.2 All emeralds are grue.

It should be clear from he meaning of the word "grue" that at time *t* all the direct evidence for H.1 is also evidence for H.2, and vice-versa. They are equally well confirmed; for at time *t* the two hypotheses have *exactly the same positive instances*. But this is paradoxical. First, the fact that all emeralds examined so far have been green, hence also grue, seems not in the least to support the prediction by way of H.2 that the next emerald examined after *t* will be blue; yet it does intuitively seem to support the prediction through H.1 that it will be green. Second, "grue" is an arbitrary predicate; there is no more reason for thinking that

[9]*Fact, Fiction, and Forecast*, Hackett Publishing Company (1973), pp. 73ff.

emeralds examined after time *t* will be blue than there is for thinking that they will be red. So there is no more reason for asserting "All emeralds are grue" than there is for asserting "All emeralds are gred." One can cook up any number of "grue"-type predicates. All will be true of emeralds to the same extent that "green" is, for the apparent hypotheses in which they figure are supported by precisely the same evidence. The evidence supports just about any similar hypothesis we wish to make about emeralds. But this result, as Goodman says, is "intolerable."

It is tempting to object immediately that "grue" and "green" are unlike in a crucial respect. The meaning of "grue" but not "green" includes a reference to a particular time *t*. Therefore, the apparent hypotheses in which they figure are not on the same footing with respect to the evidence. This objection misses the point, however. Suppose one also has available the predicate "bleen"; an object is bleen just in case it is blue before time *t* and otherwise green. Then the predicate "green" can be defined as follows: An object is green just in case it is grue prior to *t* and otherwise bleen. So, the meaning of "green" includes a reference to a particular time just as much as that of "grue." By the same token, it would be misleading to claim that an object that stayed grue would undergo an otherwise unaccountable change at time *t*. For, given its new definition, an object that stayed green would undergo the same sort of change, from grue to bleen! If there is an objection to be made here, it will have to depend on the possibility of distinguishing between *real* and *unreal* changes.

It is also tempting to claim that the "grue paradox" is a technical problem, of interest to confirmation theorists perhaps, but of little or no bearing on the actual conduct of science. This objection is not well founded. An example should make clear why not.[10]

Graphs are often used to represent statistical data and to indicate extrapolations from the data. On the next page is a fictitious graph plotting the increased use of electricity over time. Suppose use is measured at the end of every year. At the end of the first year, it measures 5,000,000 KWH, at the end of the second year 10,000,000 KWH. This information is entered on the graph as points A and B respectively. Now suppose one is asked to extrapolate the data; that is, to predict the use of electricity at the end of the third year. The problem, of course, is to find a regularity in the data already gathered and then to

[10]Compare Brian Skyrms, *Choice and Chance*, second edition, Dickenson (1966), pp. 68-71.

generalize on it. The natural thing would be to generalize on A and B by drawing a straight line through them to C and thus to predict that at the end of the third year consumption of electricity will reach 15,000,000 KWH (at the end of the fourth year 20,000,000 KWH, etc.).

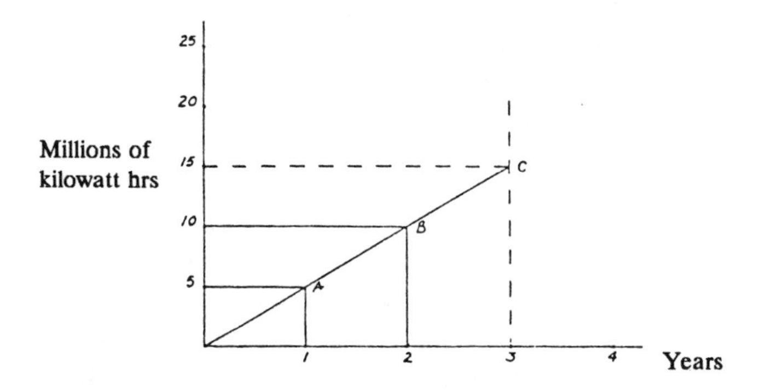

The difficulty is that this hypothesis, the interpolating between points representing data already gathered and extrapolating beyond them, is only one of a number of possible hypotheses. Here are three such possible hypotheses:

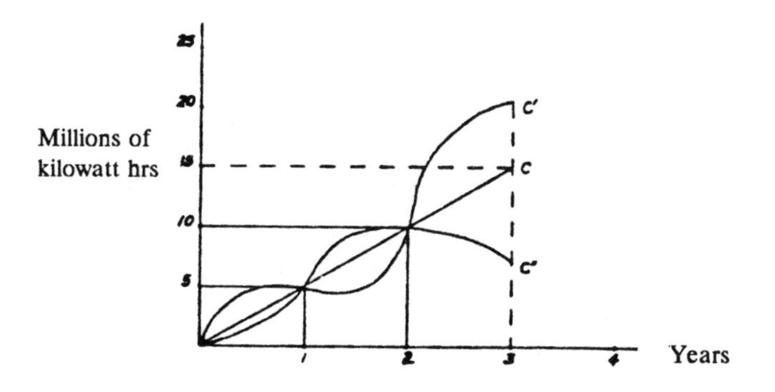

Each gives rise to a very different prediction of the use of electricity at the end of the third year, although all three agree for the data points A and B. Through any finite set of points an infinite number of curves can be drawn, each representing a possible hypothesis on the data. But this apparently implies that for any prediction one wishes to make, one can find a well supported hypothesis which licenses it. In Goodman's words, "To say that valid predictions are those based on past regularities, without being able to say *which* regularities, is thus quite pointless.

Regularities are where you find them and you can find them anywhere."[11] The "grue paradox" merely puts the problem in extremely sharp relief by showing that for any hypothesis based on the past course of our experience there is another one based on exactly the same experience which is incompatible with the first. There is not even a logical limit to the hypotheses one can manufacture.

A great number of interesting "solutions" to the "grue paradox" have been advanced, including abandoning the "positive instance" account of confirmation that generates it. Goodman's solution is to distinguish between *confirmable* and *non-confirmable* generalizations. Only the former constitute legitimate scientific hypotheses.

More specifically, Goodman holds that the second hypothesis, H.2, is illegitimate insofar as "grue" is not a *projectible* predicate. It is not projectible in the sense that the regular occurrence of emeralds and grueness together does not license the prediction or "projection" of future cases. The fact that all emeralds examined so far have been grue (because they have also been green) does not entitle us to hypothesize that *all* emeralds (past, present, and future) are grue. The task, then, is to mark a distinction between projectible and non-projectible predicates and in particular to find some asymmetry between "green" and "grue."

Given the interdefinability of "green" and "grue," Goodman infers that whether or not a predicate is projectible depends on the linguistic resources we happen to have at hand. We have already seen that "All emeralds are green" and "All emeralds are grue" cannot be distinguished on the basis of their logical form; both are universal generalizations. Nor can they be contrasted in virtue of their content, by some semantic difference in the definitions of "grue" and "green." They differ rather, according to Goodman, with respect to the use made in the past of the words "green" and "grue" in the formulation and testing of hypotheses. To distinguish between projectible and non-projectible (more accurately, between *more* or *less* projectible predicates),

> ...we must consult the record of past projections of the two predicates. Plainly 'green', as a veteran of earlier and many more projections than 'grue', has the more impressive biography. The predicate 'green', we may say, is much better *entrenched* than the predicate 'grue'.[12]

The key to Goodman's solution to his "paradox" is the concept of *entrenchment*. A predicate is projectible to the extent that it, or a

[11]*Fact, Fiction and Forecast*, p. 82.
[12]*Ibid.* p. 95.

predicate coextensive with it,[13] is entrenched in a language, and it is entrenched in a language to the extent that it has been projected by users of that language. "Grue" has not been projected by us; it has not been used to construct hypotheses licensing predictions about as yet unexamined objects. So it is not entrenched in our language (although it might have been). Hence, to complete the argument, H.2 does not provide us with a legitimate ground for prediction.

"Entrenchment" is not to be equated with "familiarity." Goodman's suggestion does not rule out such relatively unfamiliar predicates as "conducts electricity" or "is radioactive," to use his example. Entrenchment is a function not of frequency of *use*, but rather of frequency of *projection*. "is talkative," for example, is often used to describe particular individuals, but the number of times it has actually been projected, in the formulation and testing of hypotheses, is small. Despite its familiarity, it is not very well entrenched. Unfamiliar predicates, on the other hand, may be admitted in the absence of a record of past use if particular projections of them do not conflict with the projections of better entrenched predicates. "Grue" conflicts with "green," for example, but it is not easy to see what better entrenched predicate conflicts with "conducts electricity."

Goodman's resolution of the "grue" paradox has at least this much going for it: It properly calls attention to the language-dependency of the confirmation relation and it places the larger issues in an interesting evolutionary perspective. Whether an object, or even more strikingly, a fact, provides evidence for an hypothesis will depend on how the object is described, how the fact is stated, how the hypothesis is framed, and these in turn depend on the language beng used. It comes as no surprise to those involved that to do scientific work is to learn, and on occasion to extend, a particular language. In an evolutionary perspective, the entrenchment of predicates will be determined by the success or failure we, as communities, enjoy in predicting on the basis of the hypotheses in which they figure. Projectible predicates, in this perspective, are simply those that have *survived* past projections. Moreover, these two points come together insofar as there is every reason to think that our language generally has been shaped in the same way by environmental pressure and requirements of survival.

[13]Two predicates are co-extensive when they are true of all and only the same objects. For example, "man" and "featherless biped" are co-extensive predicates. There is a difficulty here, however. "Mass" came to replace "weight" in the 17th century although (a) "weight" was better entrenched, and (b) the two predicates are not strictly co-extensive.

There are, however, at least three problems with Goodman's resolution of the "paradox." In the first place, "entrenchment" is not a very precise notion. There seems to be no clear way to determine, even in otherwise well-defined cases, which of two predicates is better entrenched. In the second place, it seems to imply an overly conservative policy with respect to practice and to undermine the possibility of progress. Take, for instance, Einstein's replacement of the Newtonian concept of mass by $m = m_o\sqrt{(1\text{-}v^2/c^2)}$ where *c* is the speed of light, and his subsequent modification of Newton's principle of forces, viz., "Every object is such that the force on it equals its rest mass times its acceleration". Surely the concept of (Newtonian) rest mass was better entrenched when Einstein first promulgated the Special Theory of Relativity. This fact would seem to lead Goodman's criterion to the implausible result that Einstein's modification of Newton's principle of forces was either not confirmable or, at best, less confirmable than its predecessor. In the third place, Goodman possibly exaggerates the language-dependency of the confirmation relation. For he grants that we could, if only our language were different, have projected "grue." But then our predictions about the course of nature would have been very different. This suggest that any piece of evidence can, subject to our linguistic resources, confirm any number of incompatible hypotheses. And this is to fly in the face of the intuition with which we began, that some hypotheses are supported by the evidence for them while others are not. If one insists that there are many ways to make a world, it can only be at the price of claiming that a confirmation theory has no real work to perform.

The second "paradox" of confirmation, due to Hempel, is generally known as the "raven paradox." As already noted, it raises questions concerning what should count as a *positive* (or *negative*) *instance* of a hypothesis.

Suppose one seeks to confirm the following hypothesis:

H. All ravens are black.

On the "positive instance" account, one looks around for ravens; they constitute appropriate instances of the hypothesis, and are positive if black. But *H* is logically equivalent to another hypothesis.

*H**. All non-black things are non-ravens.

So anything which is neither black nor a raven is a positive instance (hence confirms) *H**. If it is required, as seems entirely reasonable, that objects confirming an hypothesis confirm all hypotheses logically equivalent to it—the "equivalence condition"—there emerges the paradoxical result that whatever is not black and not a raven, from white shoes to red herrings, also confirms the hypothesis that all ravens are black.

Matters are even worse. *H* is logically equivalent not only to *H** but also to *H***:

> If anything is a raven or is not a raven, then if it's a raven, it's black.

Since *H*** is logically equivalent to *H*, then given the equivalence condition, non-ravens and black objects, as well as black ravens, confirm the hypothesis that all ravens are black. The feeling of discomfiture that these examples provoke is well expressed by Nelson Goodman. "The prospect of being able to investigate ornithological theories without going out in the rain is so attractive that we know there must be a catch in it."[14] Yet the implausible result, that just about anything confirms the hypothesis that all ravens are black, arises from very natural assumptions.

Many different attempts have been made to resolve the "raven paradox." According to Hempel himself, the trouble lies neither in the original concept of an instance nor in the equivalence condition, but in thinking that the result to which they lead is paradoxical. In his words, "The impression of a paradoxical situation is not objectively founded; it is a psychological illusion."[15]

In Hempel's opinion, this "illusion" has two sources. One is the mistaken tendency to think that hypotheses of the form "All R's are B's" are *about R's* only, that *R's* constitute their subject matter. When this hypothesis is paraphrased into a language more perspicuous than English, in particular into the symbolism of modern quantifier logic as $(x)Rx \supset Bx)$, it becomes clear that such hypotheses really are about all objects whatsoever. For example, the hypothesis that all ravens are black, when paraphrased correctly, asserts that if *anything* is a raven it is also black. The class of instances of this generalization, positive and negative, is not restricted to ravens. We are misled, possibly by English grammar, into thinking otherwise and hence into mistakenly rejecting non-black non-ravens, black non-ravens, and so on, as perfectly good instances of this same generalization.

This consideration is reinforced by another. Hempel agrees that observations of white shoes and black ravens do not equally confirm the

[14]*Fact, Fiction, and Forecast*, p. 70. Note that Charlie, in an earlier example, turns out to be a positive instance of the generalization that all UCI students are wealthy, regardless of his financial condition.

[15]*Aspects of Scientific Explanation*, p. 18.

raven hypothesis, but, he points out, his account is entirely qualitative and makes no claims about *degrees* of confirmation. All that counts is whether an object is an *instance*, and, according to Hempel, it is enough that an object (or its description) *conform* to the hypothesis: that is, that an observation provides an instance just so long as it is consistent with the hypothesis.

The other source of the "illusion" that his concept of a positive instance has paradoxical consequences is, Hempel says, the mistaken tendency to take so-called "background information" into account. What leads one to rule out red herrings as acceptable instances of the raven hypothesis is the fact that one *already* knows that they are neither ravens nor black. Hence they do not provide *new* evidence for the hypothesis; they do not provide *added* support for it. But if any given test object were referred to as "object x" and characterized no further, then the discovery that it was neither black nor a raven would confirm the hypothesis that all ravens are black, though perhaps not to the same degree as a black raven would, even if the test object happens to be a red herring. If nothing is assumed in advance about the objects tested in connection with a hypothesis, if no additional information about them is allowed to intrude, then the paradox that non-black non-ravens confirm the hypothesis that all ravens are black "vanishes." Only when such information intrudes illegitimately are the consequences to which Hempel's concept of a positive instance and the equivalence condition lead paradoxical.

Two features of Hempel's account need to be emphasized. First, Hempel retains the concept of a positive instance and the equivalence condition, attempting instead to *explain away* their apparent conflict with other intuitions we might have about confirmation. Second, Hempel's account equates *confirmation* with *conformation*, construing as instances any objects whose description is consistent with the hypothesis in question.

But, it might be argued, both of these features are objectionable. In the first place, it is simply a fact that scientists, amply reinforced by common sense, do not count just about anything as instances of the hypotheses they are trying to confirm. Almost all that Hempel admits are generally taken to be irrelevant. If one requires of an adequate account of confirmation that it describe, at least to some extent, actual scientific practice, then any account which allows a piece of ice, for example, to count as evidence for the generalization "All sodium salts burn yellow" must be amended or rejected. In the second place, Hempel's account does not provide for the "testing" intuition on which Popper rightly insists. To test a hypothesis, recall, involves making a genuine attempt to refute it; that is, it involves the search for negative or disconfirming instances of the hypothesis. But if virtually everything turns up as a positive instance, virtually nothing constitutes a negative instance. One can examine a herring as carefully as we like (and some people would prefer not to do it at all), but there is no way in which that examination could generate a negative instance of, hence test, the hypothesis that all ravens are black. If one requires that genuine instances of a hypothesis are capable of testing it, then Hempel's concept of an instance must be rejected as inadequate.

4. The Bayesian Account of Confirmation[16]

Hempel's "positive instance" account of confirmation is threatened by the paradoxes of confirmation; neither of the solutions proposed inside that framework is convincing. So the question naturally arises: Are there other, more satisfactory accounts of confirmation?

The search for more satisfactory accounts is guided by three adequacy conditions. First, however abstract or idealized it might be, an adequate account of confirmation must fit actual scientific practice; second, it must resolve the paradoxes of confirmation; third, it must be grounded on a deep underlying intuition about when a scientific hypothesis is acceptable.

Currently "Bayesian" accounts of confirmation are much in the forefront of discussion in the philosophy of science. These are accounts which make important use of probability theory in assessing the confirmation of hypotheses. These accounts all agree that the evidence E confirms a particular hypothesis H just in case E raises the probability of H. That is, E confirms H just in case the probability of H given E, i.e., $P(H|E)$, is greater than the probability of H by itself, i.e., $P(H)$.

The great French mathematician, physicist, and philosopher Henri Poincare' (1854-1912) saw the close connection between confirmation and probability as follows: "...the physicist is often in the same position as the gambler who reckons up his chances. Every time that he reasons by induction, he more or less consciously requires the calculus of

[16]This presentation relies heavily on R. D. Rosenkrantz, *Foundations and Applications of Inductive Probability*, Ridgeview Publishing Company (1981). Other excellent expositions of the Bayesian account include Paul Horwich, *Probability and Evidence*, Cambridge University Press (1982) and R. C. Jeffrey, *The Logic of Decision*, second edition, University of Chicago Press (1983).

probabilities."[17] But what is the calculus of probabilities, and how is it applied to the confirmation of hypotheses?

The answer to these questions requires an elementary acquaintance with probability theory. Indeed, only a little elementary knowledge of logic and arithmetic is required to grasp the essentials.[18] So let's get to the task immediately.

In both science and philosophy of science symbols are introduced in the interest of clarity and rigor. Thus, what is sometimes called *logical probability* is the probability attaching to a statement S—$P(S)$—such that:

> (1) the probability of S is not less than zero; that is, $P(S) \geq 0$.
> (2) if S is a tautological statement—a statement which is true no matter what truth values its component statements, if any, have —then $P(S) = 1$.
> (3) If S and R are mutually exclusive statements—$(S \& R)$ is contradictory—then $P(S \vee R) = P(S) + P(R)$.

This definition yields some simple, but fundamental, theorems.

Negation: $P(\sim S) = 1 - P(S)$.

> Proof: Since S and $\sim S$ are mutually exclusive, that is, since $(S \& \sim S)$ is contradictory, then, by (3), $P(S \vee \sim S) = P(S) + P(\sim S)$. But, by (2), $P(S \vee \sim S) = 1$, because $(S \vee \sim S)$ is tautological. So, $P(\sim S) = 1 - P(S)$.

Upper Limit: $P(S) \leq 1$.

> Proof: Assume that $P(S) > 1$. Then, by Negation, $P(\sim S) < 0$, because there is a number n such that $P(\sim S) = 1 - n$ $(n > 1)$. But this contradicts (1). Thus, by reductio ad absurdum, $P(S) \leq 1$.

Logical Consequence: If R is a *logical consequence* of S, $(S \Rightarrow R)$, that is, if $(S \& \sim R)$ is contradictory, then $P(S) \leq P(R)$.

> Proof: When $S \Rightarrow R$, S and $\sim R$ are mutually exclusive. Then, by (3), $P(S \vee \sim R) = P(S) + P(\sim R)$. By Negation, $P(S \vee \sim R) = P(S) + 1 - P(R)$. But Upper Limit yields $P(S \vee \sim R) \leq 1$. Substituting $P(S) + 1 - P(R)$ for $P(S \vee \sim R)$ in the previous line

[17] *Science and Hypothesis*, Dover (1952), pp. 183-84.

[18] In what follows "p & q" is to be read "p and q", "p v q" as "p or q", and "~p" as "not p". The arithmetical symbols "+", "-", and ">" retain their usual meanings.

yields $P(S) = 1 - P(R) \leq 1$. Manipulation of the inequalities then yields $P(S) \leq P(R)$.

Logical Equivalence: if S is *logically equivalent* to R, $(S \Leftrightarrow R)$, that is, if R is a logical consequence of S and vice-versa, then $P(S) = P(R)$.

Proof: Since $S \Rightarrow R$, then by Logical Consequence, $P(S) \leq P(R)$. Similarly, since $R \Rightarrow S$, then $P(R) \leq P(S)$. Hence, by arithmetic, $P(S) = P(R)$.

Interest in the probability of some hypothesis H on the evidence E requires the concept of *conditional probability*. This concept may be explained symbolically as follows: $P(H|E) = P(H \& E)/P(E)$, so long as $P(E) \neq 0$. Now it is a crucial, if also easily derived corollary of this explanation that

$$P(H|E) = (P(E|H) \times P(H))/P(E).$$

Proof: By the explanation of conditional probability, $P(H|E) = P(H \& E)/P(E)$. Rewriting yields, $P(E \& H) = P(H|E) \times P(E)$. Similarly, $P(H \& E) = P(E|H) \times P(H)$, replacing H by E throughout. By Logical Equivalence, $P(H \& E) = P(E \& H)$. Hence, $P(H|E) \times P(E) = P(E|H) \times P(H)$. So, $P(H|E) = P(E|H) \times P(H)/P(E)$ dividing throughout by $P(E)$.

This is *Bayes' Theorem* (after Thomas Bayes, 1702-1761). It says that $P(H|E)$, the *posterior probability* of H (the new probability assigned to H) in the light of E, is the product of its *prior probability*, $P(H)$, and its likelihood, $P(E|H)$, divided by $P(E)$, called the expectedness of E. It allows one to re-adjust the probability of a hypothesis in the light of new evidence. Intuitively, this is just what a formulation of confirmation should do. Regardless of the prior probabilities a hypothesis might have, Bayes' Theorem implies that as evidence accumulates the posterior probabilities tend to converge. Moreover, the higher these posterior probabilities, the better confirmed the hypothesis and the more rational it becomes to accept it.

For present purposes, the Bayesian account of confirmation has two crucial corollaries:

C.1 Hypotheses are confirmed by their consequences. Or, more accurately, hypotheses are never disconfirmed by their consequences.

Proof: Let E be a logical consequence of H; then $P(E|H) = 1$. Then Bayes' Theorem reduces to $P(H|E) = P(H)/P(E)$. But

$P(H)/P(E)$ is never less than $P(H)$.

C.2 Of two consequences of a hypothesis, the more improbable (or unexpected) confirms it more strongly.

Proof: If E and F are consequences of H, then $P(H|E) > P(H|F)$ if $P(E) < P(F)$.

These two corollaries harmonize with important intuitions about confirmation. The first is that the (observational) consequences *deduced* from a hypothesis directly confirm it. Indeed, this is the major feature of the widely accepted "hypothetico-deductive" model of theory-testing. The other intuition is that consequences deduced from a hypothesis which are otherwise unexpected (given the presumed background knowledge and the perspective of alternative theories) tend to confirm it all the more. This was one of the main points Popper made against traditional accounts of confirmation; it motivated his emphasis on the *content* of theories and was the key to his characterization of a *test*. The intuition has often been decisive in the history of science. For example,[19] the French physicist Denis Poisson (1781-1840) deduced from the wave theory of light the observational consequence that the shadow of a small circular disk produced by a narrow beam of light should have a bright spot in its center. He was trying to falsify the wave theory; the consequence was completely unlikely on any competing theory or against the background of traditional observations, indeed so unlikely that Poisson and other members of the French Academy of Sciences thought that the wave theory had been refuted. Thus when Arago announced that the bright spot had been observed in very carefully controlled experiments, the effect was dramatic and the wave theory received quick and widespread credibility. By the same token, it follows from Bayes' Theorem that the expected consequences of a theory, those consequences likely to happen even if the theory is not true, confirm the theory much less.

The Bayesian account is quantitative. It affords the means to determine the *degree* to which new evidence confirms hypotheses. In contrast to the "positive instance account", it provides a more complex analysis of the process of confirmation because it considers the background information and theoretical contexts that alone make possible the assignment of prior probabilities, likelihoods, and expections. So the Bayesian account

[19]See Ronald Giere, *Understanding Scientific Reasoning*, second edition, p. 126.

is grounded on certain basic intuitions. The question now before the house is: How does the probabilistic account fare with the paradoxes?[20]

On the Bayesian account, it does *not* follow that they are always confirmed by their positive instances. The mere logical compatibility of evidence with a particular hypothesis is not enough to insure that the former confirm the latter. Consider, for example, the distribution of hats at random to their owners after a party. If the hypothesis is that no man receives his own hat, then the hypothesis is confirmed by the observation that two of the men received *each other's hat*. This is a "positive instance," hence for Hempel confirms the hypothesis. But if there are only three men altogether who are getting hats, then not only does the fact that two men receive each other's hat *not* confirm the hypothesis, it actually excludes it. That is, there are cases in which positive instances of a hypothesis disconfirm it. Thus the "positive instance" account must be rejected.[21] But if it is given up, then both paradoxes of confirmation are nipped in the bud. White shoes and red herrings *are* non-black non-ravens, but it doesn't follow on the Bayesian account that their observation confirms the hypothesis that all ravens are black. Similarly, a green emerald examined before time *t* is a grue emerald, but on the Bayesian account it doesn't follow that a green emerald *confirms* the hypothesis that all emeralds are grue. As the Bayesian account insists, whether the evidence confirms a hypothesis is heavily dependent on background information.

A point made earlier in the discussion of the "raven paradox" is that the "positive instance" account, because of its emphasis merely on enumeration of instances, rules out consideration of any background information—including the fact that there are many more non-black, non-ravens than there are black ravens. This fact, indeed, is an important source of the feeling of paradox in Hempel's admission of white shoes and red herrings as perfectly good evidence for the hypothesis that all ravens are black. The Bayesian account, on the other hand, easily accommodates the asymmetry. For it follows immediately from Bayes' Theorem that if the probability of non-black, non-ravens is greater than the probability of black ravens, the observation of the latter provides

[20]See Rosenkrantz, *Foundations and Applications of Inductive Probability,*, chapters 3 and 7, whose discussion we largely follow, and Patrick Suppes, "A Bayesian Approach to the Paradoxes of Confirmation," in Hintikka and Suppes, eds., *Aspects of Inductive Logic*, North-Holland (1966).

[21]For additional reasons from a Bayesian point of view for giving up the "positive instance" account see Horwich, *Probability and Evidence*, pp. 54-63.

more confirmation of the raven hypothesis than does observation of the former.

Bayesians reject the idea that "positive instances" in general and without further qualification provide confirming evidence for hypotheses. So too they include background information (for example, information about the size, composition, and description of the reference classes from which the samples are drawn). Between them these two points go some way toward disarming the "raven paradox." The first point tells against the "grue paradox" as well, for it, in turn, depends on the "positive instance" account of confirmation. In their further analysis of the paradox, however, Bayesians divide. Some, following Goodman, try to draw a sharp line between confirmable and non-confirmable generalizations (or between "projectible" and "non-projectible" predicates). One way in which this might be done is to identify confirmable hypotheses as those which have relatively high prior probabilities,[22] non-confirmable hypotheses as those which have relatively low prior probabilities, although this approach leaves us with the problem of determining the prior probabilities of hypotheses. We will turn to this problem shortly. Other Bayesians, notably Rosenkrantz in *Foundations and Applications of Inductive Probability* (chapter 7), reject the attempt to draw a sharp line between confirmable and non-confirmable generalizations (or between "projectible" and "non-projectible" predicates). They make two points in this connection. First, once Goodman's "new riddle of induction" is generalized, as in the electricity consumption example, and seen not to depend on queer predicates like "grue", it seems unlikely that a distinction between confirmable and non-confirmable generalizations can be drawn, even in principle. Moreover, there is no *reason* to do so if the only purpose is to rescue the already inadequate "positive instance" account from the paradoxes to which it leads. Second, it would be a methodological mistake simply to rule out "grue"-*type* hypotheses. We can call such hypotheses "bent"; generally, they imply theoretically well-founded deviations from otherwise "straight" hypotheses at extreme ranges of relevant variables. The "grue hypothesis" is, of course, entirely arbitrary and not theoretically well-founded. But other hypotheses having the same sort of "bent" character are. It has been proposed several times, for example, that Newton's law of universal gravitation be amended; the force between two objects varies inversely as the *square* of the distance between them, up to some distance *d*. But for distances greater than *d*,

[22]*Ibid.*, pp. 69-72.

the force varies inversely as the *cube* of the distance between them. Such proposals were made by the Royal Astronomer G.B. Airy (1801-1892) in the attempt to explain observed irregularities in the motion of the planet Uranus, and again by the 20th century physicist Seeliger in the attempt to make the mean density of the universe everywhere constant. (The inverse square law implies a concentration of matter around centers of maximum density). These proposals were not accepted, but not because they were in the intended sense "queer" or non-confirmable. Moreover, just as they postulate a spatial discontinuity in the regularities governing the universe, so the "grue hypothesis" postulates a similar temporal discontinuity. From *this* angle, there is nothing to choose between them. To ban such hypotheses from science on *a priori* grounds would at the very least bar the possibility of certain kinds of progress.

The Bayesian account of confirmation, as has been noted, neatly captures the intuitions connected with the hypothetico-deductive model of theory-testing[23] and Popper's emphasis on improbable consequences. Finally, the larger picture is appealing: Evidence confirms hypotheses as it raises their probabilities in a certain way over time; the more probable such hypotheses become, as the evidence for them accumulates, the more they are to be accepted. Two conditions of adequacy have been satisfied. What about the third, that an adequate account of confirmation must fit actual scientific practice? Most of the objections to the Bayesian account center around this question.[24]

Using Bayes' Theorem, the calculation of the *posterior* probability of some hypotheses H, given some piece of evidence E, depends on *prior* values for $P(H)$, $P(H|E)$, and $P(E)$. But what do these prior probabilities measure, and how are they to be calculated? Otherwise put, Bayes' Theorem provides a way of determining new probabilities on the basis of old probabilities. But it does not provide a way for determining the old or prior probabilities themselves. We have to find *them* in some other way. This is often termed "the problem of the priors" (i.e., prior

[23]The hypothetico-deductive model can be described briefly as follows. A "why?"-question is posed. A hypothesis is formulated in the attempt to provide an answer. Empirical implications are deduced from the hypothesis. Observations are made to check these implications. If the observations match the deductive implications, then the hypothesis is confirmed and eventually accepted. If the observations made do not match the implications deduced from the theory, then the hypothesis is disconfirmed. Just as Bayesians insist, on this model hypotheses are tested by deducing consequences from them and then seeing whether these consequences match our observations.

[24]See Clark Glymour, *Theory and Evidence*, Princeton University Press (1980), chapter III ("Why I Am Not a Bayesian").

probabilities). It suggests that at the very least the Bayesian approach is not self-sufficient and must be supplemented. A variety of proposals for understanding prior probabilities have been advanced. Among the most interesting is the proposal that they measure our initial *degree of belief* in the truth of the hypothesis under consideration[25]. To accept this proposal, however, would be to admit certain thoroughly *subjective* elements into an account of confirmation. It has also been argued that the prior probabilities represent simple generalizations on our past experience. But this last proposal appears to amount to a re-introduction of confirmation by enumeration, the "positive instance" account together with its problems. One might wonder at this point whether the Baysian approach has merely postponed rather than solved any difficulties.

Although differing greatly in their attitudes about the problem of determining prior probabilities, most Bayesians agree on two main lines of defense against this criticism. The first is that since scientific activity is never context-free, it is a merit of, and not an objection to, the application of Bayes' Theorem that it makes such context-dependency explicit. The context is formed on the basis of our past experiences and involves certain expectations about what we will observe and certain assumptions as to which hypotheses are plausible. Although admittedly in an idealized way, numerical measures can be assigned to these expectations about evidence and these plausibility estimations of hypotheses. These are what prior probabilities measure. If (what is never in fact the case) a particular experiment were context-free, no expectations about its outcomes had been formed, and our state of ignorance with respect to its prior probabilities was complete, one could assign indifferently .5 as a measure of both hypothesis and outcome prior to performing the experiment. The second line of defense is this: Whatever the initial probabilities, however widely the values assigned to them diverge, so long as they are neither 0 nor 1, the posterior probabilities will *converge* over time as the result of applying Bayes' Theorem to new evidence.

This defense, though provocative, nevertheless is difficult to reconcile, except in an extremly idealized way, with science as actually practised. In the first place, probabilistic arguments, until comparatively recently, are rare in the history of science. Only on occasion does anything like an application of Bayes' Theorem in the work of theorists or experimenters

[25]The classical (somewhat technical) exposition of this proposal is in Bruno de Finetti's "Foresight: its logical laws, its subjective sources" in Kyburg and Smokler, eds., *Studies in Subjective Probability*, John Wiley (1964).

arise, let alone any attempt, however rough, to determine the prior probabilities of hypotheses or the expectedness of outcomes. In the second place, there are several reasons why this should be so. One has to do with the role of hypotheses. We have so far presupposed a distinction between theoretical statements of hypotheses and statements describing actual or possible observations. On the Bayesian view, the second set of statements is at least as probable as the first set; in general, hypotheses will be less probable than their observational consequences. If high probability is the goal of science, then it would seem to be good policy to reduce hypotheses, wherever possible, to their observational basis. But this conflicts with the actual process of science, in which hypotheses are developed which go far beyond their observational basis and can be used to explain and predict other kinds of phenomena. And its conflicts with the accompanying intuition, clear in Popper's view, that hypotheses of *high content* should be sought, that is, theories that are highly falsifiable, for theories that are highly falsifiable can be put to genuine tests. But hypotheses that are highly falsifiable have low probabilities with respect to the already available evidence. By the same token, when a radically new hypothesis is introduced (a recurrent characteristic of the history of science and a notable engine of its progress), then typically it will have a very low prior probability in comparison with its more firmly entrenched competitors. But in that case even a positive test result confers very little additional probability on it.[26]

The other reason why scientists have apparently failed to adopt a Bayesian methodology concerns the nature of the evidence brought to bear on hypotheses. Many theories come to be accepted not because they yield novel predictions that are subsequently verified, but because they account more successfully then their competitors for observations long since made. Copernicus, for example, supported his theory with observations dating back to Ptolemy. And Newton's main case for universal gravitation was his derivation of the laws of planetary motion already established by Kepler. And so on. But in the Bayesian scheme, this sort of "old" evidence does not confirm new hypotheses, a fact which seems to make hash of the history of science. For suppose[27] that evidence E is already known when hypothesis H is introduced at time t.

[26]When Max Planck (1858-1947) introduced the notion of *quanta* to explain the observed frequencies of black-body radiation, he thought the probability of his own hypothesis was close to zero, so paradoxical did the idea of physical discontinuities seem.

[27]Following Glymour, *Theory and Evidence*, p. 86.

If E is known, then $P(E) = 1$. But if $P(E) = 1$, then the likelihood of E given H, $P(E|H)$, is also 1. Thus by Bayes' Theorem, $P(H|E) = (P(H) \times 1/1 = P(H)$. That is, the posterior probability of H given E is the *same* as the prior probability of H; E does not raise the probability of H, hence (contrary to practice and intuition) does not confirm it.

These objections do not settle the matter; Baysians have replied in detail to all of them.[28] But they do motivate the search for an account of confirmation closer to practice, one which would, if possible, retain the virtues of Baysian and "positive instance" accounts, while not suffering from their vices.

5. The "Bootstrap" Account of Confirmation

An account of confirmation intended to be more faithful to actual scientific practice than either of those discussed has been developed by Clark Glymour in his book *Theory and Evidence*. In Glymour's view, the actual practice of science is concerned mainly with establishing the relevance of a particular piece of evidence to an individual hypothesis. Often the arguments on the basis of which this is done are ingenious and complicated, as is the design and execution of experiments that put these arguments into practice. The task of confirmation theory shifts to making clear the general structure of these arguments. This in turn requires saying, in a precise way, how evidence generally supports hypotheses.

Glymour seeks not so much to supplant as to compliment either a "positive instance" or a Bayesian account.[29] This can be understood in two ways. In one way, Glymour simply adds further constraints to the previous accounts of the confirmation relation. In the other way, he proposes a strategy (or methodology) for proceeding, using one or the other of the previous accounts, to actually confirm a hypothesis.

The main idea in Glymour's account is that hypotheses are confirmed by verifying instances of them, and instances are verified by computing or determining the value of *every* quantity or property that they contain. Those quantities or properties not directly measured in, or determined by, experimental observations may be calculated by using auxiliary

[28]Replies, and an extended discussion of Glymour's position, are in John Earman, ed., *Testing Scientific Theories*, volume X of *Minnesota Studies in the Philosophy of Science*, University of Minnesota Press (1983). Horwich, *Probability and Evidence*, also discusses and eventually rejects each of Glymour's objections against the Bayesian account.

[29]In his book, his attention is directed to the "positive instance" account, but he has also shown how it fits with the Bayesian account in "Bootstraps and Probabilities," *Journal of Philosophy*, LXVII (1980), pp. 691-99.

hypotheses, including the very hypothesis from which the instance was derived. Quantities that are neither measurable nor computable are indeterminate, and hypotheses containing them connot be tested; such hypotheses are empirically meaningless. To be more precise, hypotheses are confirmed by the evidence with respect to a theory just in case:

> (i) a "positive instance" of the hypothesis (in Hempel's sense) can be deduced from the evidence with the help of theory ("confirmation clause");
> (ii) other evidence might have led to the derivation of a "negative instance" ("non-trivialization clause");
> (iii) the theory may include the very hypothesis under consideration ("bootstrap clause").

Confirmation, then, is a three-termed relation linking a piece of evidence E to a hypothsis H by way of a background theory T, with the proviso that H might be included in T. It is this last feature, that a hypothesis may be assumed in the course of testing it, that leads Glymour to claim that a hypothesis can and often does support (or "lift") itself by its own bootstraps.

In the typical case, the testing strategy is as follows. An experiment to test a hypothesis is carried out against a particular theoretical background. The experiment involves making a series of measurements. Some quantities that the hypothesis ascribes can be determined by the experiment; others cannot. Those that cannot (often some theoretical constant) are computed by means of other hypotheses of the theory (or even the original hypothesis itself). The computed values confirm or disconfirm the hypothesis by way of providing positive or negative instances of it.

To illustrate, one cannot test Newton's Second Law of Motion, force equals mass times acceleration or $F = ma$, by measuring m and a and using the hypothesis to calculate F, for there is no other way to determine F experimentally than by measuring m and a. But one can use the Second Law of Motion to obtain a value of F with which, once the masses m_1 and m_2 of two objects and the distance between them R have been measured, to test Newton's Law of Universal Gravitation, $F = G(m_1m_2/R^2)$. In this case, one part of Newton's theory is used to derive values used in testing another part of the theory.

To illustrate a properly "bootstrapping" case, consider the testing of the Ideal Gas Law.

> Hypothesis: for any sample of gas, other things being equal, the product of the pressure and volume of the gas are proportional

> to the temperature of the gas, $PV = kT$, where k is an undetermined constant.

Suppose P, V, and T can be measured, but not k. The hypothesis is tested by obtaining two sets of measurements of P, V, and T. Then the first set of values together with the hypothesis to be tested is used to determine a value for k: $k = PV/T$. The value for k, together with our *second* set of measurements of P, V, and T results in an instantiation of the hypothesis (the number computed by multiplying values for P and V does in fact equal the number computed by multiplying k by the value for T), in which case the hypothesis is confirmed; or it does not result in an instantiation of the hypothesis, in which case the hypothesis is not confirmed. The crucial points are that a value for k had to be determined before the hypothesis could be tested, since a hypothesis can be tested only when *all* its values are determined, and that one, although perhaps not the only, way to do this is to use the hypothesis itself. The apparent circularity involved is not vicious, for the second set of measurements *could* have led to a negative instance, that is, values such that $PV \neq kT$. The "bootstrap" strategy requires that in every case such negative instances are at least possible.

The situation is usually much more complicated than this. Usually several different quantities or properties are experimentally or observationally undetermined and a number of theoretical assumptions are necessary to calculate them.

Here is another example, somewhat simplified, which concerns that landmark of qualitative biology, the theory of natural selection.[30]

> Hypothesis: the key to differentiation (at least among birds) is geographical isolation.

This hypothesis is an important part of the theory of natural selection. One piece of evidence for it is provided by the 13 species of Darwin's Finches in the Galapagos Islands (together with the one species on Cocos Island to the northwest), first spotted by Darwin on the *Beagle* voyage of 1835. If obviously differentiated from a common ancestor (a fact which can be supported independently) in terms of color, body size,

[30]Based on David Lack's article, "Darwin's Finches," *Scientific American*, 1953. It is interesting that Lack, a great ornithologist, seems aware of the "bootstrap" character of his argument, and of the hypothetical connections on which it rests, for he says that "Darwin's finches provide *circumstantial* evidence for the origin of a new species by means of geographical isolation" (our italics) and yet this "circumstantial evidence" was among the most important supports (and the initial source) of Darwin's theory.

and primarily beak development, they constitute a partial instance of the hypothesis. But without some additional inferences, they don't constitute a complete instance. When Darwin first saw them, they were not geographically isolated; several species were living in close physical proximity. If the finches are to provide an instance of, hence confirm, the hypothesis, it also has to be established that they were originally geographically isolated and only eventually interpopulated the Islands. This can be done with the aid of two additional hypotheses of evolutionary theory. One of these auxiliary hypotheses is that if populations of two sub-species "have not been isolated for long and differ in only minor ways, they may interbreed freely and so merge with each other." The other hypothesis is that if such populations have been isolated for a long period of time, "so many hereditary differences will have accumulated that their genes will not combine well. Any hybrid offspring will not survive as well as the parent types." But the various finch species, even those living side by side on the southernmost island, have not merged. So it may be concluded that they were originally geographically separated for a relatively long period of time. *Now* we have a confirming instance of our hypothesis. What saves the argument from triviality is that the auxiliary hypotheses are themselves supported by different kinds of evidence (drawn, for example, from a study of insect populations) and competing hypotheses (for example, that new species evolve by becoming adapted to different habitats in the same area) have not been confirmed.

So a case for the *fit* of the "bootstrap" account with actual scientific practice can be made. To what intuitions does it appeal? In the first place, it preserves the "sampling" and "test" intuitions. The "non-trivialization clause," in particular, ensures that the hypothesis being tested is always at risk; the strategy is so rigged that positive outcomes are not guaranteed. In the second place, and more importantly, the "bootstrap" account appeals to what might be termed a "selective support" intuition.

On several accounts of confirmation, apparently including both (unrevised) Bayesian and hypothetico-deductive accounts, if a hypothesis H is confirmed, then so is H conjoined with any other hypothesis G, no matter whether the evidence at hand bears on G or not, just so long as H and G are consistent with each other and with the evidence E. But, intuitively, such irrelevant conjuctions should not be taken as confirmed. A great part of scientific work, as Glymour insists, consists of devising experiments that test *specific* hypotheses within a theory, as against testing the theory as a whole. Which hypotheses are tested by which pieces of evidence depends on the structure of the argument linking them. Consider Kepler's Laws of Planetary Motion.

1. Each planet moves in an elliptical path with the sun at one focus.
2. The line joining the sun and the planet sweeps out equal areas in equal times.
3. For any two planets, the ratios of the squares of their orbital periods (T^2) is equal to the ratio of the cubes of their average distances from the sun (R^3).

The first and second laws have to do with the motion of individual planets around the sun. The third has to do with the ratio of the periods of any two planets. To test the first and second laws, we make observations of a single planet; suitably chosen, four such observations suffice. To test the third law, however, we need observations of two planets. Thus evidence for the first two laws is not at the same time evidence for the third, nor does support accrue to the third even if we conjoin it with the first two.[31]

"Bootstrap" testing gives us a particularly clear way of distinguishing between hypotheses tested and untested (in an irrelevant conjunction, for instance) on particular occasions. It is implied by Glymour's account that confirmation of a hypothsesis involves derivation of instances of every quantity or property in it. Thus, observations might support the hypothsesis that $A + B = C$, but not the conjunction $A = B + C$ and $C = D \times E$, and hence not $A = B + (D \times E)$, because values for D and E might not be computable from these observations.

One final point concerning our intuitions about confirmation might be made. It certainly doesn't follow from, but in some sense is required by, the "bootstrap" account that the more *varied* the evidence for a hypothesis the better. At the very least it draws attention, in a way that the others do not, to this important point. For on the "bootstrap" account, it is particularly clear that a variety of evidence is needed to allow us to separate hypotheses. If one hypothesis is confimed by making computations from the evidence using another hypothesis, it is always possible, other things being equal, that both hypotheses are false; errors in one are possibly compensated by errors in the other. The only way to guard against such errors, as in the evolutionary example, is to test the hypotheses in as many different ways as possible, and this requires a

[31]It should be noted that on the Bayesian account observation of a single planet may confirm the conjunction of Kepler's third law with the first two, although it doesn't follow from this that observation of a single planet can confirm Kepler's third law in isolation.

variety of evidence. As Glymour points out, in the 17th century confirmations of Kepler's First Law of Planetary Motion required assuming the Second Law, and confirmations of Second Law required assuming the First Law. Many astronomers, in fact, were unclear whether errors in one were compensated by errors in the other. Not until the invention of the micrometer and Flamsteed's (1646-1719) observations of Jupiter and its satellites could the Second Law be confirmed without using the First Law.

We have looked at the first two standards of adequacy, "fit" with actual scientific practice and "intuitive appeal." It remains to say something about the "paradoxes of confirmation" before turning finally to some difficulties with the "bootstrap" account.

First, the "raven paradox." As has been noted, Glymour retains the idea that hypotheses are confirmed by their positive instances, and to this extent his "bootstrap" account is vulnerable to the difficulties raised by the "raven paradox." But Glymour thinks that these difficulties are mitigated if not entirely resolved by the following consideration.[32] On Hempel's account, a white shoe provides an instance of the generalization that all ravens are black, but it also provides an instance of the generalization that all ravens are not black. But these generalizations are *contraries*: If there exists a raven, then they cannot both be true. A white shoe provides no basis for distinguishing between them, whereas a black raven is an instance of the first generalization and not of the second. If one held that a genuine instance should provide a basis for distinguishing between them, then one could conclude that a white shoe was not a genuine instance of the generalization that all ravens are black, and hence did not provide evidence for it. The problem is to elaborate the "positive instance" account in such a way that only black ravens are retained as *genuine* positive instances.[33]

The "bootstrap" strategy provides us with one means for doing so, at least in certain kinds of standard contexts. On the "bootstrap" strategy, instances are derived from hypotheses only with the help of some background theory. But the background theory typically will discriminate evidence more finely than will the hypothesis itself and allow us to restrict the class of potential instances.

[32]See Richard Grandy, "Some Comments on Confirmation and Selective Confirmation," *Philosophical Studies*, 18, (1967), pp. 19-24.

[33]This will not resolve all the difficulties for there are cases (recall the exchange of hats between three men after a party) when even *genuine* positive instances do not confirm their corresponding generalizations.

For example, suppose that *Ra & Ba* confirms a hypothesis of universal conditional form with respect to some background theory *T*. It doesn't follow, as it would on Hempel's original "positive instance" account, that *~Ra & ~Ba* also confirms it with respect to the same theory *T*. For example, if the hypothesis is $(x)(Cx \supset Dx)$ and the background theory is $(x)(Rx \supset Cx) \& (x)(Dx \equiv Bx)$, then *Ra & Ba* confirms the hypothesis, but *~Ra & ~Ba* does not; for from *Ra & Ba* we can deduce a positive instance of the hypothesis via the background theory, while from *~Ra & ~Ba* we cannot deduce either a positive or a negative instance of it. *Ra & ~Ba* would at the same time allow us to deduce a negative instance. But this is just the hoped-for result. Against the background of theory *T*, black ravens confirm, non-black ravens disconfirm, and white shoes do neither.

Second, the "grue paradox." Goodman originally intended it to show that confirmation is not a purely syntactic relation between sentences. If $(x)(Ex \supset Gx)$ is the hypothesis and *Ea and Ga* is a sentence describing evidence, then if "E" means "is an emerald" and "G" means "is green," then, the evidence sentence is true, and intuitively confirms the hypothesis. But if "E" means "is an emerald" and "G" means "is grue," then the evidence sentence is true, and intuitively does *not* confirm the hypothesis. It all depends on how you *interpret* the symbols or syntax. Goodman drew the conclusion, we saw earlier, that confirmation is relative to a particular interpretation or, equivalently, to a given language. Glymour believes that this conclusion can be extended to any formal confirmation theory, whether it be the "positive instance," Bayesian, or his own "bootstrap" account.[34] Take for example, a probabilistic condition on confirmation: *E* confirms *H* just in case $P(H/E) > P(H)$. Let *H* be $(x)(Ex \supset Gx)$ and *E* be *Ea & Ga*. Then again on the first ("green") interpretation, *E* will intuitively confirm *H*, but on the second ("grue") interpretation intuitively *E* will not confirm *H*. And so on, Glymour believes, for every formal confirmation theory of the type we have been considering. The moral, apparently, is that the "grue paradox" cannot be used to decide between existing confirmation theories. None of them can resolve it without also adding certain presuppostions about meaning. Thus failure to do so should not be held against the "bootstrap" account. A formal confirmation theory applies only if the interpretation of the language to which it is applies is restricted.

[34]As expressed in correspondence with the authors.

Finally, we must mention what seems to be an important difficulty with the "bootstrap" account. Recall that it was so framed as to fit actual scientific practice, in particular to fit the often difficult testing of individual hypotheses by peculiarly *relevant* pieces of evidence. But it is not at all clear that the "bootstrap" account manages to solve all of the problems that it intends to solve in connection with relevance. To the contrary, it underlines one of the traditional problems in a striking way.[35] This is the problem of *ad hoc* hypotheses. *Ad hoc* ("for the purpose") hypotheses are especially concocted to explain data that would otherwise disconfirm a theory but have no further explanatory value. *Ad hoc* hypotheses are irrelevant to the genuine explanation of data and the testing of theories. The problem is to build a criterion of relevance into one's account of confirmation in such a way that they can be ruled out. The "bootstrap" account fails to do so.

The central feature of the "bootstrap" account is that evidence is related to a hypothesis by way of auxiliary hypotheses and background theories. The difficulty is that with a suitably chosen auxiliary hypothesis one can relate *any* piece of evidence to *any* hypothesis. Thus whatever sort of object $Fa \& Ga$ might describe, and whatever hypothesis $(x)(Rx \supset Bx)$ might express, to relate the former to the latter all we need do is assume some auxiliary hypothesis of the form $(x)(Fx \supset Rx) \& (x)(Gx \equiv Bx)$. Now of course this won't do. One cannot simply assume just *any* auxiliary hypothesis in the course of testing scientific theories. Only the *right sort* of auxiliary hypotheses may be assumed, i.e., ones that are not *ad hoc*. But what is the "right sort" of auxiliary hypothesis? Glymour suggests that it is part of some *accepted theory* or that is has been *supported by a variety of evidence* in a number of different ways. This, however, seem more to postpone than to resolve the difficulty. For if we ask on what grounds the background theory has been *accepted* or in what ways the auxiliary hypotheses have been *supported*, a criterion is still needed indicating why the theory was worthy of acceptance and it can only be one which shows why its support was not vacuous. If no criterion is forthcoming, then there is no reason on the "bootstrap" account for preferring some theories to others. The reason is that, without such a criterion, any two theories can always be confirmed by whatever evidence is available so long as one is ingenious enough to find a connecting hypothesis. In other words, the "bootstrap" account seems to lead either to an infinite regress or to the holistic claim that all of science (that is,

[35]See Daniel Garber, "Old Evidence and Logical Omniscience in Bayesian Confirmation Theory," in *Testing Scientific Theories.*

the set of all supported hypotheses) lifts itself together by its bootstraps. But this latter claim undermines the intuition at the very heart of every confirmation theory, that the evidence supports hypotheses, one at a time, in some specifiable way.

6. Assessment: The Problem of Relevance

The three theories of confirmation reviewed in this chapter provide different treatments of the relation:

a confirms *b*, where *b* is a universal generalization.

According to the Hempelian view, the relation in question holds just in case *a* is a positive instance of *b*; according to the Bayesian theory, it holds just in case the probability of *b* given *a* is greater than the probability of *b* alone; and, according to the Bootstrap theory, the relation of confirmation holds just in case *a* is a *genuine* positive instance of *b*, where what constitutes a genuine positive instance is, in part, a function of a theory T or the background knowledge K.

Hempel's "positive instance" account of confirmation accords hypotheses support in intuitively irrelevant ways, irrelevant *evidence* in the case of the "raven paradox," irrelevant *hypothesis* in the case of the "grue paradox." The Bayesian account of confirmation excludes evidence as irrelevant that, intuitively and in practice, supports hypotheses. This is the problem of "old evidence." The "bootstrap" account of confirmation, though going some way toward resolving the "raven paradox," and admitting "old evidence," has some relevance problems of its own. The problems of confirmation have, it seems, converged on a central issue, the problem of relevance.

Chapter four

THEORIES

I. Introduction

The discussions of explanation and confirmation have alluded to a distinction between *theories* and *experimental "data"*, although we have said very little about either theories or experiments. The distinction is so well entrenched that most people think of scientific activity as divided into two parts, theorizing and experimenting, and many scientists describe themselves as primarily involved in one or the other.

The questions raised by this distinction are of more than academic interest. Contemporary opponents of the classical Darwinian account of the origin and differentiation of species, for instance, variously claim that evolution is no more than a "theory" (as opposed to a fact), that it is not a "*scientific* theory" (because, it is claimed, evolutionary hypotheses are not in principle falsifiable) and that it is not a "*well-confirmed* theory" (because, so "scientific creationists" say, there are too many experimental facts for which it cannot account). These anti-evolutionary claims cannot be understood, much less settled, in the absence of a clear understanding of what theories are, how they are related to facts, and why we choose some in preference to others.

We will look at three different views of scientific theories, guided in each case by the kinds of question raised in the evolution-creation debate. We will also be interested, however, in two further questions, answers to which do not directly emerge from the three views to be examined:

> How seriously should we take the concepts and objects that different theories introduce? (are they mere "constructs" or do they give us the truth about the constituents of the world?)
> How is the "growth of scientific knowledge" or "progress" to be characterized?

These two questions have always been at the center of discussion about the nature of scientific theories.

First, recall that the problem of confirmation arises because generalizations "go beyond" their supporting evidence; an empirical generalization such as "All ravens are black" may be false even though every raven *so far* observed is black. *Theories* "go beyond" the evidence in a much more radical way as well. They often appeal to unusual, and in many respects unobservable, entities such as positrons and neutrinos, forces and fields, drives and motives. Indeed, in the case of theories, more than mere confirmation or disconfirmation typically is at stake; in the case of theories questions about the *empirical significance* of the postulation of theoretical objects arise (that is, about how such objects relate to the experimental "data") as well as questions about their *reality*. Do theoretical objects exist or is talk about them merely a convenient way of organizing and interpreting sequences of observations? If scientific methodology demands that what one scientist demonstrates another can duplicate, how are claims about atoms or the unconscious to be interpreted, let alone decided? These questions bother some scientists as much as they bother philosophers because most contemporary scientific theories make liberal use of terms supposedly referring to unobservables.

Second, there is the fact of scientific change. It is the major fact with which we began this book. Viewed from a distance, the history of science is one long procession of theories, one succeeding another over the course of time. But there have also been at least two scientific "revolutions," one in the 16th and 17th centuries, the other at the end of the 19th and the beginning of the 20th centuries, in which everything that had gone before seemed suddenly to be repudiated and rendered false. As a result, many scientists have adopted a cautious and pragmatic attitude about the theories with which they work. Once burned, twice afraid. Even Newton's physics, dominant for over 200 years, was apparently overturned by Einstein. Is there any reason to think that any of our *present* theories will not some day be similarly overturned? More generally is it possible to find a pattern in the flux of historical change?

Has it been *rational* in some intrinsic way? Is there progress in the history of science as well as procession?

2. The Classical View of Theories[1]

A view of theories that implies answers to some of the questions raised in the previous section and suggests answers to others takes the distinctive features of theories to be features of their *linguistic formulations*. It holds that a scientific theory is a set of sentences formulated in a language with a clearly specified vocabulary and structure. These sentences are of two basic types. There are, first, the *theoretical principles* of the theory, typically mathematically phrased, expressing relations between various kinds of things; "$PV = rT$," "$R = V/I$," "$s = 1/2gt^2$" are examples. There are, second, "*correspondence rules*" which provide the theoretical principles with empirical content by linking at least some of the symbols in the principles to empirical determinations. Thus, the symbol "P" in the first of our examples could be linked by a correspondence rule to an empirical determination of *pressure*, "R" in the second of our examples to an empirical determination of *resistance*, "t" in the third of our examples to an empirical determination of *time*. Typically what is meant by an "empirical determination" is simply some way of *measuring* the quantity in question; "t", for example, is determined by the use of clocks.

Running hand in hand with the distinction between types of *sentences* is a distinction between types of phrases or *terms*. Terms are either *theoretical* or *observational*. "Has a mean kinetic energy E" and "is repressed" are examples of the former, "is red" and "grinds his teeth" are examples of the latter. Correspondence rules serve as a means by which theoretical terms are given observational (or measurable) meaning. Most early versions of the classical view conceived of correspondence rules as *definitions*; each theoretical term was meaningful to the extent that it could be defined in observational terms. Operationalism—the influential view that every scientific term is synonymous with "the set of operations" by which it is determined or measured—is possibly the most extreme of

[1]This view is also known as the "received view" of theories. But the view is no longer standard and the word "received" does not indicate that it is an element in the logical empiricist position in the philosophy of science. There is an excellent introduction, both historical and philosophical, to most of the issues and much of the literature surrounding it in Frederick Suppe, *The Structure of Scientific Theories*, second edition, University of Illinois Press (1977).

these versions.[2] The difficulty with operationalism is that it implies that every time we come up with a new method of measuring, say, mass or length or habit strength, we thereby give a new meaning to "mass" and "length" and "habit strength." But this implies that these terms are ambiguous, whereas, intuitively, we think that they have only *one* sense each, however many different measuring operations are used in their application.[3] So proponents of the classical view have generally given up the possibility of *explicitly defining* theoretical terms in an observational vocabulary, preferring to say that each operation on the basis of which we apply a theoretical term only *partially* specifies its meaning.[4] But whatever form the correspondence rules take, the important points are (a) that a sharp distinction between theoretical and observational terms is assumed and (b) that correspondence rules of some sort are needed to provide the former with meaning in terms of the latter. It is only by way of such rules that theoretical principles find *application* to our experience; it is only by way of them that we can derive the testable implications of these principles.

N. R. Campbell, one of the founders of the classical view of theories,[5] uses the suggestive terms "hypothesis" and "dictionary" to refer to the two basic types of sentences that comprise a theory. The "hypotheses" are both the *axioms* of the theory—the fundamental theoretical principles—and the *theorems* following from them—the non-fundamental theoretical principles. The "dictionary" connects these hypotheses with physical reality by providing at least some of their constituent "hypothetical" terms with empirically determinable meaning.

Campbell constructs the following rather abstract and somewhat fanciful example to illustrate his position:

[2]The classic statement of the operationalist position is in the Nobel laureate P. W. Bridgman's *The Logic of Modern Physics*, Macmillan (1927). The concept of an "operational definition," frequently encountered in the behavioral and social sciences, derives from Bridgman's position, although it is now often used in a rather loose and off-hand way.

[3]For a critique of the operationalist position, see C. G. Hempel, "A Logical Appraisal of Operationism," reprinted in *Aspects of Scientific Explanation*.

[4]The "partial interpretation" view of the meaning of theoretical terms is worked out by Rudolf Carnap in "Testability and Meaning," *Philosophy of Science*, 3 (1936), pp. 428-68, and 4 (1937), pp. 1-40. Excerpts from this very important paper have been widely reprinted, for example, in Feigl and Brodbeck, *Readings in the Philosophy of Science*. Carnap focusses his discussion on the problem of defining so-called disposition terms such as "soluble," a problem that turns out to be well-nigh insoluble.

[5]In *Foundations of Science*, Dover Publications, (1952). Originally published as *Physics: the Elements* (1919).

I. hypotheses

(i) u, v, w, ... are independent variables.

(ii) a is a constant for all values of the variables,

(iii) b is a constant for all values of these variables.

(iv) $c = d$, where c and d are dependent variables.

II. dictionary

(i) the assertion that $a(c^2 + d^2) = R$, where R is a positive and rational number, implies the assertion that the resistance of some definite piece of pure metal is R.

(ii) the assertion that $cd/b = T$ implies the assertion that the temperature of the same piece of pure metal is T.[6]

From the hypotheses we can deduce, by purely mathematical reasoning, that

$$a(c^2 + d^2)\, cd/b = 2ab = \text{constant.}$$[7]

When this proposition is interpreted by means of the dictionary, we obtain

> the ratio of the resistance of a piece of pure metal to its absolute temperature is constant,

which purports to be an empirical generalization.

Campbell's view has two corollaries. One is that the theory (hypotheses + dictionary) enables explanation of the purported generalization by showing how it follows from the basic principles of the theory. The other corollary is that the purported generalization helps to provide a way of testing the theory because it has obvious observational consequences.

Those who accept the classical view of theories are not concerned whether it accurately describes the structure of actual scientific theories. It is instead a *philosophical reconstruction* that makes both the empirical and non-empirical features of a theory clear. It is irrelevant that most scientific theories don't explicitly display two basic types of sentences, theoretical principles and correspondence rules. The point is that they

[6]We have followed Campbell's way of phrasing the dictionary entries. In fact, his way seems to run together two different purposes which definitions serve, (a) to introduce a new symbol, in order to abbreviate and abridge (e.g., $R = a(c^2 + d^2)$, $T = cd/b$) and (b) to provide empirical criteria on the basis of which to apply the symbol introduced (e.g., "R" designates the resistance of some piece of metal, "T" its temperature, both resistance and temperature to be measured in the usual ways).

[7]Since $c = d$, $(c^2 + d^2) = 2c^2$ and $cd = c^2$; hence $2ac^2\, c^2/b = 2ab$. Since a and b are constants, $2ab$ = constant.

could, and that reconstructing theories in this syntactic way allows us to answer most of our initial questions.

In the first place, the misleading distinction between theories and facts is replaced by a clear distinction between theoretical and observational sentences. The distinction between theories and facts is misleading because it suggests that "theories" are not well established and that "facts" are. But as scientists use the word, calling a set of explanatory principles a "theory" has nothing to do with the degree to which they have been established. Some theories are very well established (including the theory of evolution[8]); others are not. The distinction between theoretical and observational sentences, on the other hand, is well-founded because theories are introduced to explain what we observe and what we observe in turn confirms certain of our theories. The merit of the classical view is that it makes the relations between these two types of sentences clear and precise.

In the second place, the classical view of theories shows how theories, even those employing such intuitively non-observational concepts as "force" and "field" and "the unconscious" have empirical significance. On this view, a theory has empirical significance *only if it has testable consequences*. It is one of the principle functions of correspondence rules to establish a bridge between theoretical principles and observational fact. Thus a theory which employs the concept of "the unconscious" is empirically significant only if it has testable consequences in the form, for example, of predictions about the way people with certain past histories will *behave* in certain situations, a connection presumably effected with the help of correspondence rules. The real challenge for "scientific creationists" is not to find phenomena which the theory of evolution allegedly cannot explain, but to show that their own theory has testable consequences. For testability, on the classical view, is the way in which science and theology are to be distinguished.

These two points have an important corollary. It is that we can always settle (at least in principle) the claims of rival theories by performing a *crucial experiment*. A crucial experiment involves deriving observational predictions, by way of correspondence rules, from competing theories. Since the theories are competing, the predictions must differ; at most one can be true. It follows that a crucial experiment will refute at least one of the two theories.

[8]For a convenient summary of the evidence and an interesting discussion of the evolution-creation debate, see Philip Kitcher, *Abusing Science*, MIT Press (1982).

There are many examples of crucial experiments in the history of science. One of the most famous concerned a long-standing controversy about the nature of light. There are two familiar theories about light. One is associated with the Newtonian conception of the world. It holds that light consists of very small particles moving at very high velocities. The other theory holds that light consists of waves. Now it can be inferred from the particle theory that light should travel faster in water than in air, and the precise amount can be calculated. But it can be inferred from the wave theory that light should travel slower in water than in air, and again the amount can be calculated. It was not until about 1850 that instruments were sophisticated enough to measure accurately the predicted velocities. When an experiment was at last performed, it turned out that light does travel *slower* in water, and by the amount predicted by the wave theory. The result of this experiment was general acceptance of the wave theory and general rejection of the particle theory, at least until Einstein provided new foundations for the latter in 1905.

In the third place, while the classical view does not imply an answer to questions concerning the reality of theoretical entities,[9] it does suggest a picture of the growth of scientific knowledge and underlines the rationality of scientific change. Theories, by way of correspondence rules, confront experience in the form of the observations we make. This experience represents a neutral tribunal at which the claims of various theories are to be judged. Typically the theories come to the tribunal in pairs and a crucial experiment is performed to decide between them. Thus the evaluation of theories is entirely objective. One theory displaces another according as it is better able to predict and/or explain the course of our experience. Consistent with the idea that theories come and go while their observational basis remains the same, some classical theorists have offered the following account of the growth of scientific knowledge.[10] A theory succeeds another because it is better confirmed or because as an extension of the earlier theory it now embraces more phenomena or because it consolidates heretofore separate domains. In the first case, a theory that is well confirmed with respect to the instrumentation and observations available is disconfirmed in the light of better subsequent measurements. The replacement of the Ptolemaic by

[9]Some philosophers who subscribe to the classical view are scientific realists; they think that at least some of the entities postulated by scientific theories exist. Their opponents (some of whom do espouse the classical view) are called anti-realists.
[10]See Ernest Nagel, *The Structure of Science*, chapter 11.

the Copernican theory is a case in point (since Ptolemy's theory was shown to be less accurate in its prediction of planetary positions), and perhaps the same could be said of the replacement of the particle by the wave theory of light after 1850. In the second case, a theory is supplanted by another when the first can be reduced to the second (in the sense that its theoretical principles can be derived from the theoretical principles of the second theory, sometimes with the addition of assumptions limiting its domain of application). What characterizes this type of "homogeneous" reduction is that the two theories use approximately the same concepts and the phenomena described are qualitatively similar. The extension of classical particle mechanics to the study of rigid bodies, the absorption of Galileo's laws into Newton's physics, and the reduction of Newton's theory to Einstein's (on the assumption that the speed of light is infinite) are often cited as examples. In the third case, a theory is supplanted by another when the first can be reduced to the second by way of additional assumptions correlating concepts in the first theory with concepts in the second; this is "heterogeneous" reduction. Classical thermodynamics and statistical mechanics, for example, employ quite different concepts and were originally used to describe dissimilar phenomena, but the former can ostensibly be reduced to (derived from) the latter essentially when we add to statistical mechanics a statement correlating or identifying temperature with mean kinetic energy. Similarly, it has been claimed that classical genetics reduces to molecular biology once the chemical nature of the gene has been correctly understood.

Two final points concerning the classical view help to place it in larger perspective. The first is that it embraces a "hypothetico-deductive" view of theories. In part this means that a theory contains "hypotheses" (the theoretical principles) from which certain consequences, given empirical content by the correspondence rules, can be deduced. As we have seen, the hypotheses are intended to explain and in turn to be confirmed by these observational consequences. But the view also implies that a philosophical account of scientific theories must attend strictly to the role and status of hypotheses, their function and form, and ignore the question how they came to be discovered. The historical, psychological, and social circumstances in which the theory came to be discovered (Newton watching an apple fall in his garden, for example) and advanced are irrelevant. What matters only are its conceptual structure and the ways in which it can be put to empirical test. As Hans Reichenbach phrased it, we are interested in a theory's "context of justification" not its "context of discovery."[11]

The second point to be noted is that the classical view is often associated with a belief in, and a picture of, the logical unity of the sciences. On this view, both the history and the present structure of science can be represented by a pyramid or "layer cake" of deductively linked sentences, sentences describing individual observational facts at the base or bottom, empirical generalizations on the next level up, and so on. Logic, the correspondence rules, and sometimes theoretical connections or "bridge laws" (in the case of heterogeneous theories) take us from one level to the next. The situation can be pictured as in the figure on the next page. The "higher" levels are more "general" in the sense that the lower levels can be derived from them, but not vice-versa. Thus, that a piece of chalk falls when dropped is an individual fact whose description can be derived from the empirical generalization "All unsupported objects fall."

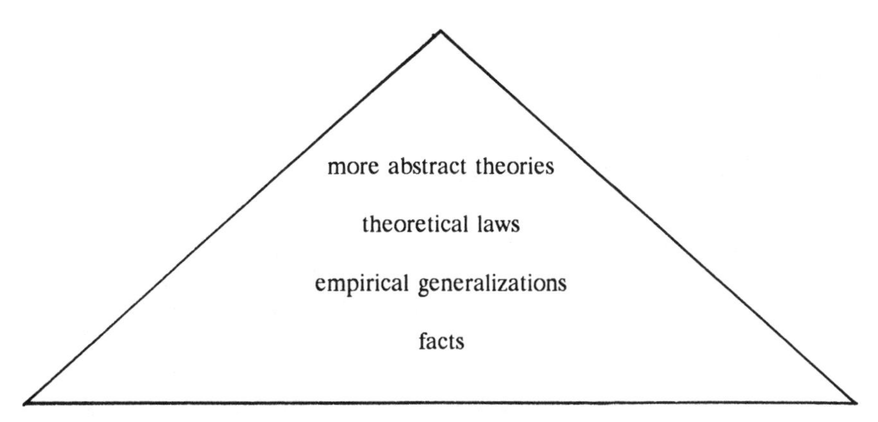

The latter follows from Newtonian gravitational principles which in turn follow from Einstein's general theory of relativity. The "layer cake" view makes clear what seem to be very plausible claims: that the progress of science has been marked by the transition to theories of greater explanatory power, that science tends towards a more unified structure, and that the whole edifice of science rests on a solid and empirical foundation.

The classical view of theories has much to recommend it. It makes the structure of theories precise, provides a way of understanding their

[11]See his *Experience and Prediction*, University of Chicago Press (1938).

explanatory power, and bridges the apparent gap between theories and observations. Most important, it entails that any genuine *scientific* theory must have empirical significance and that empirical significance is gained only if a theory has testable consequences. Theories make assertions—either true or false—about the way the world is. If these assertions are empirically significant, it must be possible to confirm or disconfirm them at least indirectly. It is in this way that theories are *objective* accounts of the phenomena.

But the classical view has also been subjected to a variety of criticisms. To better understand these criticisms, recall that the classical view has two main elements: (1) theories are to be construed as formal or linguistic structures (like arithmetic or geometry); (2) a sharp distinction can be made between "theoretical" and "observational" sentences and terms, and that it often incorporates the claim that individual facts, empirical generalizations, and theoretical hypotheses form a "cake", the layers of which are linked by logical relations and correspondence rules. These basic features of the classical view bear the brunt of the criticism.[12] We will consider each in turn.

Granting the controversial claim that every theory can, in principle, be formalized, would it in fact be philosophically interesting to do so? Two different considerations suggest that it would not. One is that any number of *different* formal organizations of the same theory can be given, all having the same deductive consequences. But the freedom to choose among axiom sets suggests that formal organization *per se* will not isolate or clarify the fundamental principles of a theory. Axiomatization is primarily an expository device; it does not allow us to see how or why certain theoretical principles have been chosen as basic. The second consideration is that axiomatization or formalization *is* important in the case of logical and mathematical theories. But the reason, it has been argued, is precisely because these theories are not empirical. The meaning of their basic concepts, and hence the truth of the propositions into which they enter, is entirely "determined" by the axioms. The truth of empirical theories, on the contrary, is determined by the way the world is rather than by the way in which they are formalized. This explains, moreover, why logicians and mathematicians are themselves

[12]In addition, some philosophers claim that the classical view is far too simple an account of scientific theories, that "its very sketchiness makes it possible to omit important properties of theories and significant distinctions that may be introduced between different theories." Patrick Suppes, "What is a Scientific Theory?" in S. Morgenbesser, ed., *Philosophy of Science Today*, Basic Books (1967).

interested in formalization, whereas physicists, biologists, and so on, rarely if ever are.

Proponents of the classical view insist on the *form* a theory takes partly because of the belief that questions concerning the origins of the theory, the intentions of the theorizer, etc., are irrelevant to the philosophical appraisal of theories, just as the same sorts of questions are irrelevant to the formal appraisal of arguments. All that counts are the fundamental assumptions and what is provable from them, the "context of justification" rather than the "context of discovery." But recent work in the history of science challenges this conviction.[13] To really understand a particular historical episode, it has been suggested, one must proceed as an anthropologist. When an anthropologist studies some unknown tribe, he or she begins to understand and correctly describe the actions of its members only when she begins to understand the intentions with which they are performed, the meaning which the members of the tribe give to them. A particular series of movements counts as a rain dance and not as a general outbreak of some nervous disease only when, for example, it has been determined that the tribe is *trying* to induce the gods to send rain. In the same way when we describe a particular development in the history of science, we do so correctly only when we describe it from the point of view of those who took part in it; only then can its real significance be gauged. The English chemist Robert Boyle's (1627-1691) definition of an element[14] *looks* very much like a *contemporary* definition, and some historians have been tempted to treat it this way. But it is a mistaken reading because, although Boyle uses *words* which resemble those used in the contemporary definition, he understands those words in a very different, characteristically 17th century, sense. To understand what *Boyle* meant, and hence to appreciate not only his historical role but the content of his theoretical assertions, it is necessary to reconstruct the context in which he lived and worked, the framework of concepts and assumptions in terms of which he described the results of his experi-

[13]See T. S. Kuhn, *The Structure of Scientific Revolutions*, second edition, University of Chicago Press (1970).

[14]"And, to prevent mistakes, I must advertize to you, that I now mean by elements, as these chymists that speak plainest do by their principles, certain primitive and simple, or perfectly unmingled bodies; which not being made of any other bodies, or of one another, are the ingredients of which all those called perfectly mixt bodies are immediately compounded, and into which they are ultimately resolved." *The Sceptical Chymist*, Everyman (1911), p. 187. For support of the claim made about Boyle's definition, see Marie Boas (Hall), *Robert Boyle and Seventeenth-Century Chemistry*, Cambridge University Press (1958), Chapter III.

ments, and his intentions.[15] But this context, of course is the "context of discovery."

The sharp distinction between theoretical and observational terms demanded by the classical view has also been challenged on several different grounds.[16] In the first place, it has never really been made precise, and it is doubtful whether it is possible to do so. Intuitively, "is round" is clearly observational and "has a relativistic mass of *n* grams" is clearly theoretical. But when one presses the distinction, or tries to formulate a criterion on the basis of which to decide individual cases, one quickly runs into trouble. Is "temperature," for instance, an observational or a theoretical term? It is not easy to say. If one requires that a term applied on the basis of instruments such as a thermometer be deemed theoretical, then "temperature" is a theoretical term. Yet at the same time "temperature" seems to be an observational term because its application does not depend directly on a particular theory. Appeal to the use of *instruments* in formulating a general criterion does not seem to be very much help in any case since the use of one sort of instrument, say a microscope, surely allows us to *observe* single-celled animals, while the use of another, say a Wilson cloud chamber, apparently affords only the *inference* that alpha particles have particular properties. Moreover, what is to count as an "instrument"—eyeglasses, a thermometer, an electron microscope, a linear accelerator—is left up in the air.

In the second place, many theoretical terms, terms often used to formulate the basic principles of a theory and whose introduction helps to organize and explain the course of our experience, seem to be just as clearly observational—at least in the sense that the objects and properties to which they refer can be observed more or less directly. "Cell," "charge," and "stimulus" are examples. Indeed, Newton, in elaborating his particle theory of light, talks about "red corpuscles," thereby applying an

[15]As Kuhn points out in *The Structure of Scientific Revolutions*, science textbooks habitually review the history of their subjects (in an introductory chapter) from the present point of view, reading significance back into the past and looking at previous efforts as steps leading up to present developments (the important figures were "precursors"). But such an approach is ahistorical and seriously distorts the meaning of various experiments and theories. Herbert Butterfield's very interesting book, *The Origins of Modern Science*, Macmillan (1960), first published in 1949, was an early and influential attempt to restore an historical perspective to the history of science.

[16]For example, by Peter Achinstein, *Concepts of Science*, Johns Hopkins Press (1968), Mary Hesse, *The Structure of Scientific Inference*, Macmillan (1974), and Hilary Putnam, "What Theories Are Not," in Nagel, Suppes, and Tarski, eds., *Logic, Methodology, and the Philosophy of Science*, Stanford University Press (1962).

observational predicate to a theoretical object, happily riding roughshod over a distinction on which the classical view rests.

The underlying difficulty with the classical view in the present regard lies in the belief that observational data are the bedrock on which theories ultimately rest. The test of any theory is whether it adequately accounts for the observational and hence pre-theoretical data. But at best this is a naive view. What counts as a datum, what we can be taken to "observe," depends on a background of interpretive theories.[17] There are, in fact, no observational data apart from particular "theories of observation," sets of rules forming part of all scientific activity that tell us how to "read the data." Many of these rules have to do with optical corrections that have to be made, with respect both to the observer and with respect to the instruments that he or she uses. A simple example is the apparent passage of the sun from east to west. Copernicus' theory tells us that this is not really something we observe; what we think we see is merely illusion, to be corrected or explained by the hypothesis that we ourselves are in motion. A more complex example concerns Newton and the first Astronomer Royal, John Flamsteed. Flamsteed's very careful work by means of telescope and clock on the precise determination of the positions of the fixed stars disclosed certain discrepancies with the data predicted on the basis of Newton's theory of celestial motion. Confronted with these discrepancies, Newton kept providing Flamsteed (to the latter's eventual irritation) with optical corrections to be made in connection with telescopic sightings and thus squared his theory with them. This suggests that a sharp distinction between theories and observations is impossible to make out. Paradoxical as it seems for those who insist on the distinction, apparent refutations of a theory on the basis of observational data are often turned into striking confirmations of it.[18]

A third sort of criticism leveled against the classical view is directed specifically at its "layer cake" component. An influential, very vigorous

[17]A convincing case can be made for the claim that the theoretical/observational distinction is itself theory-relative, i.e., what counts as observational for one theory may count as theoretical for another.

[18]See Imre Lakatos, "Falsification and the Methodology of Scientific Research Programs," in Lakatos and Musgrave, eds., *Criticism and the Growth of Knowledge*, Cambridge University Press (1970). Lakatos sets out a number of other cases that also show the impossibility of describing facts separately from particular theoretical contexts.

argument is advanced by Paul Feyerabend.[19] According to Feyerabend, two conditions are presupposed by the "layer cake" account: the *consistency condition* and the *condition of meaning invariance*. The "consistency condition" requires that any *new* theory be an extension of some previously accepted theory or, at the very least, be consistent with such previously accepted theories. The consistency condition thus incorporates the classical view that scientific change is cumulative. Although one theory may be replaced by another (the second theory might be more inclusive or more accurate than the first), the first theory, if it is well confirmed, remains a permanent part of the scientific record and is simply treated as a limiting or special case. The condition of "meaning invariance" requires that the meaning of theoretical terms does not shift as new phenomena are described and explained. Unless meaning remained constant we couldn't picture scientific change as cumulative. Presumably if these two conditions did not hold it would be impossible to derive validly, and hence on the classical account to explain, one layer of the "cake" by another. But, Feyerabend insists, these two conditions do not conform to scientific practice. They are constantly violated; their violation is in part what characterizes a scientific "revolution." In addition, the consistency condition is inherently unreasonable. First, the condition "eliminates a theory not because it is in disagreement with the *facts*; it eliminates it because it is in disagreement with *another theory*, with a theory, moreover, whose confirming instances it (sometimes) shares. It thereby makes the as yet untested part of the theory a measure of validity. But a theory should not be rejected simply because it is a newer theory." Second, the only good argument for the consistency condition rests on a mistaken view of the nature of observational data, viz., that they are autonomous and hence may be used, as the bottom layer of the cake, to provide an objective criterion against which to assess rival theories. But the observational data are not, even relatively, autonomous. Every observational datum, we saw in the last paragraph, is dependent on how the data are interpreted. But such interpretation is tied to a particular theoretical perspective. A corollary to this claim is that "new facts" often come to light only with the development of a new theoretical perspective which is inconsistent with ones already in use. Sometimes this inconsistency escapes notice. Defenders of the classical view often claim that classical or phenomenonological thermodynamics

[19]Perhaps the simplest presentation is in his paper, "Problems of Empiricism," in R. Colodny, ed., *Beyond the Edge of Certainty*, Prentice-Hall (1965). The passages following are quoted from this paper.

reduces to statistical mechanics and the kinetic theory of matter. According to Feyerabend, however, it does not reduce in a straightforward way. For the facts about the Brownian particle and its motion which refute the second law of classical thermodynamics[20] depend on assuming the perspective of statistical mechanics and the kinetic theory. But this is possible only because the latter perspective is in certain respects inconsistent with that of classical thermodynamics. To generalize on this example: If potentially refuting data are to be discovered, then the consistency condition will at least sometimes have to be violated. But for Feyerabend, violating the consistency condition, in the interests of a *truly* empirical methodology which submits theories to the greatest possible number of tests, is tantamount to giving up the classical account.

Not only is that classical view mistaken, in Feyerabend's eyes it is also pernicious. This view, he insists, is a stultifying dogma. It bars the door to future progress because it precludes the type of revolutionary development (introduction of new theories which are inconsistent with those already in place, radical shifts in the meaning of theoretical terms, and so on) on which such progress depends. At the same time, the classical attitude toward metaphysics, as consisting of systems of unverifiable and therefore empirically insignificant sentences, needs to be revised. For Feyerabend, not only is there no way to distinguish sharply between legitimate science and illegitimate metaphysics, but metaphysical systems, which are for him nothing but scientific theories in an early stage of development, deserve to be encouraged rather than philosophically demolished.

3. The Historicist View of Theories[21]

Alternatives to the classical view by and large reject a sharp distinction between theoretical and observational terms and the "layer cake" picture of scientific change. The first of these alternatives, the historicist view of theories, is based on a close study of the history of science.[22]

[20]For details see Karl Popper, "Irreversibility, or Entropy since 1905," *British Journal for the Philosophy of Science*, VIII (1957).

[21] This view of theories has also been called the "subjectivist" view. But "subjectivist" is a misleading way to characterize an account of science that emphasizes its *social* aspects. Moreover, the proponents of the current view have much in common with a philosophical movement that flourished at the beginning of this century known as historicism.

[22]See Paul Feyerabend, *Against Method*, New Left Books (1975); N. R. Hanson, *Patterns of Discovery*, Cambridge University Press (1961); T. S. Kuhn, *The Structure of Scientific Revolutions*; Stephen Toulmin, *The Philosophy of Science*, Hutchinson (1953).

The historical study of science has had two quite different implications, one having more to do with the *substance* of the history of science, the other having to do more with its *methodology*. The substantive implication is this: A careful study of the history of science reveals that it has passed through a number of conceptual *revolutions*, a fact difficult to reconcile with the claim of the classical view that the history of science is the story of the gradual accumulation of deeper and more precise information about the world and ourselves—a corollary of the "layer cake" account. Take, as a well known example, the transition from Aristotelian to Galilean physics.[23] Into what "layer cake" can both physical theories be fitted? None, apparently, because not only is Galileo's theory not an extension of Aristotle's theory, it is inconsistent with it. Galileo rejects the commonsense belief that Aristotle takes as fundamental: A net force is required both to move an object and to maintain it in motion. This conceptual difference is so basic that it undoes most comparisons of the two theories[24] and certainly the picture of them as successive stages in the construction of a "layer cake" of change. Old-fashioned history often pictured social, political, and economic change in the same way, as progressing through "logically linked" stages to some "advanced" state or goal. This picture, long since scrapped in the listed areas, apparently is no more reasonable vis-a-vis scientific change. The history of science is like any other sort of history; it is not a case of linear development.

The methodological implication has to do with the "anthropological" approach to the history of science. To study history as *history*, it is contended, is to recreate the past in its own terms and not to read the present back into it. A corollary is that episodes in the history of science can only be understood from the "inside," within the "context of discovery," from the point of view of the participants. Those who subscribe to the anthropological approach are therefore sometimes referred to as "subjectivists." They insist on the subjective significance of scientific theories, the meaning the theory has for those who advance it. This opposes the classical view's insistence on the *objective* significance of theories, something accessible to everyone by way of correspondence

[23]For an introductory account, written, however, from the point of view of the classical position, see Gerald Holtan, *Introduction to Concepts and Theories in Physical Science*, Addison-Wesley (1953).

[24]What "crucial experiment" *could* one perform that would show the one theory correct, the other not?

rules, whether or not one holds the theory or understands the intentions behind it.

To clarify further the historicist view, we shall focus on the work of Thomas Kuhn. His book *The Structure of Scientific Revolutions* has been its most influential presentation. Although Kuhn's conception of theories is rooted in a revolutionary conception of the history of science and an anthropological methodology, he emphasizes the importance of tradition and the stability of scientific communities. Science depends, in his view, not just on the insights of individuals but on the existence of communities bound together by shared concepts, assumptions, and methods. Whatever progress there is in the history of science takes place within such communities, in the research tradition of what Kuhn calls "normal science."

Kuhn makes use of three key concepts in addition to that of "normal science." They are the concepts of *revolution*, *paradigm*, and *anomaly*. A brief discussion of each will make clear the outlines of his position.

Kuhn contrasts the *revolutionary* with the *evolutionary* character of scientific change emphasized by the classical view. The former involves "the community's rejection of one time-honored scientific theory in favor of another incompatible with it" (p. 6). Moreover, the political analogy contained in the word "revolutionary" draws attention to the various stages of revolutionary development. Indeed, for Kuhn science is essentially a social, if not also a political, activity and is to be understood in social terms and in a social context. For the most part such activity is conservative; revolutions break out only when it has become impossible to repress the dissidents. In other words, revolutions are rare.

The characterization of "normal science" is largely sociological. It means "research firmly based on one or more past scientific achievements that some particular scientific community acknowledges for a time as supplying the foundation for its further practice" (p. 10). It's what most scientists do most of the time. Kuhn accepts much of the classical view when it is relativized to a particular tradition of normal science. Within such a tradition one can identify such typical stages as observation, hypothesis formation, prediction, and verification, all roughly as the classical view describes them. What the classical view cannot do is to describe or account for abnormality, those sharp shifts in point of view that constitute a revolution. Such shifts aside, most scientific work consists in working theories out, solving puzzles, within a very limited range of possible and guaranteed solutions.

The notion of a *paradigm* is more difficult. Kuhn characterizes it in a variety of ways, but in none is it clear precisely what the relation between "paradigm" and "normal science" is, except for the fact that traditions of normal science take place within particular paradigms. The notion is important because if one knew how to characterize paradigms, then a *revolution* could be understood just as a shift from one paradigm to another.

One might think of a paradigm as a *theoretical framework*, provided that "theoretical framework" is widened to include sample problems, techniques and methods, instrumentation, etc., as well as the hypotheses or theoretical principles one usually associates with such a framework. But the notion of a "theoretical framework" thus widened still does not allow one to make either precise or plausible in an obvious way two of Kuhn's central claims: that *paradigms determine facts* and that *paradigms are incommensurable*. It is more instructive to take paradigms as *languages*.[25]

That paradigms determine facts can be understood in at least two ways. The "weak" interpretation is that paradigms limit the range of *appropriate facts*; they determine what sort of evidence is relevant for a theory and the classes of cases it is intended to explain. But the classical theorist could admit as much and continue to maintain a sharp observational/theoretical distinction. The "strong" interpretation is that paradigms determine not what facts are appropriate (where we should direct our attention), but what the facts *are*; there are no facts without theories or, as it is sometimes put, all facts are theory-laden. The strong claim can be explained via the suggestion that paradigms are languages as follows. A fact is a state of affairs described in a certain way. But then it seems to follow that paradigms determine facts because what the facts are will depend on what language is used to describe them; two distinct descriptions, two facts—in other words, no two meaningfully distinct descriptions can be given of the *same* fact. Paradigms determine facts in the sense that what the facts are depends on the linguistic resources available.

There are greater difficulties in interpreting the claim that paradigms are *incommensurable*, all of them having to do with the vagueness of "incommensurability." It cannot mean simply that different paradigms lead to different predictions because then pairs of them could be

[25]Although for Kuhn paradigms are non-linguistic, he seems to agree that the main argument for incommensurability rests on an identification of paradigms with languages. See his "Reflections on My Critics," in Lakatos and Musgrave, eds., *Criticism and the Growth of Knowledqe*, Cambridge University Press (1970), a collection of papers on various aspects of Kuhn's position.

confronted with the observational facts, and crucial experiments performed. But Kuhn wants to deny that they can be compared in this way. More plausibly, two paradigms, considered as languages, are "incommensurable" when there is no translation of one language into the other.

An argument for incommensurability can be developed in terms of the anthropological approach suggested earlier.[26] An anthropologist comes upon a tribe speaking a language wholly unknown to him. He tries to correlate the sounds the tribesmen make with objects in his environment. Since "gavagai," for example, is uttered on all and only those occasions on which a rabbit is present, the anthropologist correlates "gavagai" with the English word "rabbit." In this way, inductively, the anthropologist develops a translation manual on the basis of such matchings. But, even with this translation manual in hand, the relation between the two languages will remain indeterminate in the sense that although the two languages are empirically equivalent (the native makes certain sounds on all and only the same occasions that the anthropologist uses certain English words), it is nevertheless possible that the anthropologist has mistranslated the native's utterances. More generally, to put the point in terms of the classical conception of theories, theoretical sentences are underdetermined by the totality of all observation sentences; two theories can differ even if they share the same observation sentences. I might translate "gavagai" as "rabbit," when in fact it refers to part of a rabbit.[27]

Now think of the Galilean and Aristotelean paradigms as two languages. Speakers of the one need not understand speakers of the other, even though they use the same words, "pendulum" and "falling stone" (or, "earth-seeking body"), on all and only the same occasions. For calling something a pendulum is not simply to describe a falling stone in another vocabulary; it is also to presuppose a number of theoretical commitments, hence to implicate a paradigm. Furthermore, Kuhn stresses, what we observe is never simply *given*, but is always a function of expectations and past experiences. There is no important sense of "observe" in which people operating with different paradigms observe the same thing. There is no common ground of experience against which to compare theories with each other. The double failure to find common

[26]The argument derives from W. V. Quine, *Word and Object*, Wiley and Sons (1960), who deploys it for different purposes.

[27]In fact, anthropologists took the aboriginal "kanga-roo" to refer to the animal invariably present when they pointed quizzically in its direction, when in the native's tongue "kanga-roo" means "I don't know."

ground in the language of paradigms (since there is no way on the basis of experience to guarantee the adequacy of translation from one paradigm to the other) or in the observational experience to which terms of that language refer (since the fact that what we see is what we look for is in large part determined by the paradigms with which we operate) implies that theories are incommensurable or incomparable. We can't translate theories into theories, or facts into facts. How, then, if paradigms can't be directly compared with each other or with pre-theoretical or theory-neutral observations, can one paradigm ever come to replace another and what role do experiments play? To answer the first of these questions, Kuhn brings the key concept *anomaly* into play. Anomalies are sets of data that reigning paradigms have difficulty in accommodating. Of course, there are always expedient ways of accommodating them, either by explaining them away (as mentioned in the chapter on explanation) or by adopting one or another *ad hoc* hypothesis (as mentioned in the chapter on confirmation). Sometimes this tactic is successful. At other times it is not; a paradigm becomes so overburdened with *ad hoc* hypotheses and attempts to explain away otherwise disconfirming observations that it begins to sink of its own weight and scientists begin casting about for new and simpler paradigms.

Kuhn provides a more extended discussion of the role and purpose of experiments. Experiments play three different sorts of role. In the first place, experiments are carried out to determine more accurately central paradigmatic constants, e.g., the boiling points and acidity of solutions, the speed of light, etc. In the second place, a "smaller class of factual determinations is directed to those facts that, though often without much intrinsic interest, can be compared directly with predictions from the paradigm theory." On the classical view, of course, this is the most important role that experiments play. Kuhn plays it down. Such experiments rarely if ever provide a *test* of a theory, for one thing because the paradigm theory is itself invariably implicated in the design of the apparatus used for the experiments; the theory only "tests" itself. In the third place, a most important class of experiments involves attempts to *articulate* the paradigm, that is, to solve problems that involve its application or extension. But again, the working out of a paradigm via such experiments presupposes the paradigm and does not really test it.

Theories or paradigms are, in Kuhn's phrase, "ways of seeing the world." There is no way in which they can be compared with each other (because they are mutually incommensurable or untranslatable) or with experience (because the observational facts are themselves theory-laden).

It follows that three central claims of the classical view of theories—that a sharp theoretical/observational distinction can be drawn, that theories can be evaluated in an entirely objective way, and that scientific knowledge is cumulative (one generation, as Newton put it, seeing farther because it stands on the shoulders of its predecessors)—must be given up.

The historicist view first came to attention in the 1960's, challenging old and settled beliefs in a way characteristic of that generation. But, in the perspective of time, it has been subjected to serious criticism. All of these have to do in one way or another with the rather sweeping *relativism* to which it leads. The historicist view holds that the questions we put to nature as well as the appropriate answers to them, even the observational facts themselves, are determined by particular paradigmatic theoretical contexts and that these contexts cannot be compared or evaluated. But this is very implausible.

First, the historicist view conflicts with the very historical *facts* with which it began: that there have been important *changes* in the history of science and that one must *adopt the point of view of a theory* (or paradigm) before one can understand it. The historicist, it seems, must find it a complete mystery why such change occurred in the first place. It is not open to the historicist to say that one theory replaces another because it is now better confirmed or because as an extension of the earlier theory it now embraces more phenomena or because it consolidates heretofore separate domains. But it itself supplies no other persuasive explanation. For example, it should be clear that on Kuhn's account the emergence of anomalies is not enough to *force* a paradigm shift. One can always accommodate such anomalies by adding *ad hoc* hypotheses or dismiss them as not needing explanation, much in the same way that Copernicans simply dismissed the fact that objects did not fly off a moving earth as an objection to their hypothesis. Indeed Kuhn resorts to images and analogies, invoking "gestalt switches" and the like, admitting (pp. 92-3) that paradigm shift is more a question of persuasion than of nature or logic.

Not only is scientific change a mystery on the historicist view, but, the criticism continues, it fails to provide a plausible motive for carrying out experiments. If the results of every scientific investigation can be reconciled with any given theory, then experiments would seem to have a limited role at best—certainly nothing like the central role the classical view assigns to them. Kuhn does downplay the importance of experiments. But this conflicts with the historical facts that incredible patience and subtlety have been expended on experiments designed to *test* individual theories and that the results of such experiments have often been taken as "crucial" in deciding between rival theories.[28] Indeed, if the actual intentions of scientists are taken into consideration, as proponents of the historicist position urge, inevitably we will be led to emphasize the testing role of experiments because this is how most scientists have conceived of them.

Finally, there are very general difficulties with the type of relativism, theoretical or cultural, which stems all too naturally from taking what we have called the "anthropological" approach.[29] In various ways it is incoherent. For one thing, the sketch Kuhn gives us of various paradigms —pre-revolutionary Aristotelian and post-revolutionary Galilean—presupposes as a condition of our understanding the differences between them that we understand them both. Contrary to what Kuhn and the historicists suggest, we can and must use a common language to talk about varying paradigms; to the extent that we succeed in doing so the ground beneath the claim that paradigms are not intertranslatable is undermined. For another thing, there is something paradoxical about the historicist metaphor of differing perspectives or points of view. The metaphor makes sense only if there is some way we can compare them, say as different locations on the same landscape. But again the possibility of such comparison is explicitly ruled out on the historicist view. Finally, meaningful differences between theories are possible only when there is some common ground on which to compare them; taking paradigmatic theories as languages, this comes to saying that we must understand one another. In sum, the historicist case undercuts itself, by denying through its relativism the very assumption (that speakers of different languages can understand one another, even in circumstances of radical translation) that makes the statement of that relativism coherent in the first place.

[28]In addition to the examples already given in the text, we might mention that scientists themselves thought Lavoisier's experiments with combustion decided the issue between the phlogiston theory and its critics, just as a later generation believed that the famous eclipse expeditions of 1919 to determine the planet Mercury's orbit precisely tipped the balance in favor of Einstein's general theory of relativity.

[29]What follows is our very brief gloss on Donald Davidson's important and complex paper, "The Very Idea of a Conceptual Scheme," reprinted in *Inquiries into Truth and Interpretation*, Clarendon Press (1984).

4. The Semantic View of Theories[30]

The principal difficulty with the classical view is its indefensible distinction between theoretical and observational terms. The principal difficulty with the historicist view is a sweeping relativism that is ultimately incoherent. Hence the problem is to reconcile a non-relativist or objective account of theories with giving up the theoretical/observational distinction. The semantic view purports to be such a view.

On the classical view theories (construed as sets of sentences) are identified with their formulations and, as a result, the fundamental questions about theories are about their linguistic aspects. Historicists deny that theories are to be identified with their formulations, but the only plausible arguments for their claims that observations are theory-laden and that theories are incommensurable seem to depend on taking paradigms as languages. But the identification is unhelpful for at least three reasons.[31] In the first place, if theories are identified with their formulations, then each new formulation yields a new theory. But our common intuition is that the *same* theory can be formulated in a number of different ways and in different languages. In the second place, it is doubtful that a distinction between types of terms, between theoretical and observational terms, can serve to isolate the empirical content of a theory. Two points might be made in this connection. One is that many "theoretical" terms can be characterized in an "observational" vocabulary simply by negating "observational" predicates; for example, a physical field is unlike a physical object in that it is not solid, not colored, etc. The other point is that science itself, and not philosophers engaged in conceptual analysis, determines the limits of what is observable, in part by way of theories about human perception and physiology. In the third place, the fundamental questions about theories are not linguistic in character. Questions about inter-translatability, in particular, shed little light on the role and status of theoretical entities or the rationale of scientific change.

[30]See Ronald Giere, *Understanding Scientific Reasoning*, second edition, Chapter 5; Fredrick Suppe, *The Structure of Scientific Theories*, second edition, pp. 221-30; Patrick Suppes, "What is a Scientific Theory?" in S. Morgenbesser, ed., *Philosophy of Science Today*, Basic Books (1967); Bas van Fraassen, *The Scientific Imaqe*, Chapter 3. Giere's account is perhaps the most accessible; we have drawn on it in what follows.

[31]Van Fraassen, *The Scientific Image*, p. 56, impugns the entire formalist program: "The main lesson of twentieth-century philosophy of science may well be this: no concept which is essentially language dependent has any philosophical importance at all."

So, some philosophers have argued, we should abandon the focus on the sentences that formulate a theory and concentrate instead on what makes these sentences true, its *semantics*. This approach involves the introduction of a new concept, the concept of a *model*. This concept is already in wide use among scientists; what used to be called "theory-construction" is now called "model-building." This fact alone provides ample motive for taking a closer look at the semantic view and at the concept of a model that it employs.

A *model* is a representation of a physical system such that certain relational properties of the physical system are preserved in the representation, i.e., the model is structurally similar or "isomorphic" to the system (at least to its empirically determinable part). In many sciences the representation ideally is quantitative, i.e., the model is a "mathematical model," and predictions concerning the behavior of the physical system can be made on the basis of the representation and then checked out in the laboratory. Good models are confirmed in the course of such checking; they are then said to "fit" the data. Equally important, they suggest new and unexpected experiments as well.[32]

Perhaps the most familiar of all theoretical models is the molecular model of a gas, the molecules represented by elastic spheres which, within a closed space, have a mass and motion and are subject to the basic laws of statistical mechanics. In terms of this simple model we can explain such various phenomena as the mobility and mutual mixing of gases. We can also explain the observed relations between the pressure, volume and temperature of a gas as codified in the Boyle-Charles Law. In this example, a model is an idealized replica of a physical system.

More specifically, a theoretical model posits a set of objects whose properties and behavior are characterized by certain general laws.[33] For example, a Newtonian particle system is a system (model) of particles (point-masses) that satisfies (makes true) Newton's three laws of motion together with the law of universal gravitation. Since a model is a system of objects with respect to which the sentences of a theory are true, no matter how these sentences happen to be formulated, different formula-

[32]This is to characterize "models" slightly more narrowly than logicians and mathematicians do. Common to all uses of "model" is the idea that a model is a set of entities—physical objects, points, numbers, whatever—together with a set of relations between them such that the sentences of the theory modeled are true. Thus, the natural numbers and the relations into which they enter provide a model of the usual axioms of arithmetic.

[33]In the molecular model of a gas, these laws are the laws of statistical mechanics; by way of the model they serve to explain such phenomenological regularities as the Boyle-Charles Law.

tions (of the axioms of Euclidean geometry, for instance) have the same models. Models are non-linguistic structures.

To get clearer about the sense of "model" at stake here, let us contrast theoretical models with scale and analog models.[34] A *scale* model is of course a physical replica of some object or system, often much smaller than the original (model airplanes), although occasionally much larger (models of the parts of the body used as illustrations in biology classes). Scale models are sometimes used not simply to illustrate but also to explain a particular scientific phenomenon. Perhaps the most famous example is the double helix model of DNA that Watson and Crick constructed out of wire and metal in the 1950's which allowed them to understand its chemical structure. Theoretical models are like scale models in that they depend on a similarity of structure with what they model or represent, but they are unlike scale models in that they do not consist of miniaturized or enlarged physical representations of the phenomena. Mathematical or computer models are *structurally* similar to certain kinds of phenomena, but they don't look like them at all. A computer model can be designed to simulate the weather (most predictions are now made on the basis of such simulations), but this doesn't at all depend on having it rain inside the central processing unit. All that is important is that the model be like the phenomena in certain respects; given certain inputs, it can provide outputs which, given an empirical interpretation, match those actually observed.

Theoretical models are also like *analog* models. An analog model depends on an analogy between some familiar or well understood phenomenon and one whose main characteristics have yet to be discovered. Analog models, too, sometimes figure in the development of scientific theories. Thus, Rutherford and Bohr very much clarified the structure of the atom when they provided the familiar "solar system" model of it (electrons orbiting around a central nucleus). An analog model is like a theoretical model in that, once again, both depend on a similarity of structure. But they are unlike in that (a) a theoretical model is not necessarily more familiar or well understood than the phenomenon it represents (although the laws that govern it are or can be made more precise), and (b) a theoretical model does not depend on an obvious (in the case of the Rutherford-Bohr model, an obvious *visual*) analogy. The only "analogy" required is that the data curves the model implies in fact "fit" the curves actually calculated.

[34]See Giere, *Understanding Scientific Reasoning*, second edition, pp. 78-81.

Since the distinction has become so important, we should add a word about *deterministic* and *stochastic* models.[35] The contrast depends on the concept of a *state of the system*. The state of a system is its complete characterization, in terms provided by the appropriate theory, at any given moment. One can, for example, specify the state of a Newtonian system by giving the positions, masses, and momenta of all the particles that are included in it at any given time. Now a deterministic model is one in which the state of the system at any one time completely and invariably determines its state at all other times. Given a complete description of the system at any one time, one can infer a complete description of the system at any other time. A Newtonian particle system is a classic example of a deterministic system. A stochastic system, on the other hand, is such that the state at any one time determines no more than the *probability* of various states at other times. Theoretical models in genetics are typically stochastic: The characteristics possessed by parents do not invariably determine the characteristics possessed by their offspring (we can only infer the possibility of a child's having a certain characteristic on the basis of the fact that one or both of the parents have it).[36]

According to the semantic view of theories, then, theories consist of theoretical models in addition to the *empirical hypothesis* that the model(s) approximate parts of the real world in certain respects. To the extent that the approximation is good, i.e., the model "fits" the data, the theory explains why things happen as they do and the data confirm the theory.

Now it can be shown how the semantic view handles the difficulties lodged against its classical and historicist rivals.

In the first place, as already stressed models are not linguistic. The same model or class of models can be described in a number of different ways and in many different languages. The various axiomatizations of particle mechanics vary, but they all define or specify the same class of models.

In the second place, the distinction between model and data is not a distinction between what is "theoretical" and what is "observational." The model typically employs terms that one might classify indifferently as theoretical and observational in its description. Thus the description of

[35]We will come back to this distinction in the final chapter on the limits of scientific explanation.

[36]For a very interesting application of the semantic view to the study of genetic theories, see Elizabeth Lloyd, "A Semantic Approach to the Structure of Population Genetics," *Philosophy of Science*, 51 (1984), pp. 242-64.

a Newtonian particle system uses both the intuitively theoretical term "point-mass" and the observational term "position." Similarly, the data are described in ways that mix theoretical and observational vocabularies (on the assumption that the distinction has any content at all). The data we gather in connection with the molecular model of gases, for example, often require very sophisticated measuring instruments and depend on the application of certain highly theoretical concepts (e.g., "equilibrium state"). It is a corollary of these points, moreover, that the semantic view does not assume that *the data* are simply "given," apart from any and all theoretical contexts.[37] Data are always data for a given theory; all that is required is that they can be described in such a way that their "fit" with the theory isn't guaranteed in advance. The data for a Newtonian particle system or model are described in exactly the same terms as is the system, but this doesn't preclude the possibility that these data might not fit or confirm it.

In the third place, the semantic view provides for the *objectivity* of theories. For every theory includes an empirical hypothesis to the effect that the model in question does in fact apply to the data intended. This hypothesis is either true or false and can, at least in most cases,[38] be checked out. If there is a good fit, if parts of the model can be confirmed independently, if a variety of evidence has been brought to bear, and so on, then we can say that the empirical hypothesis to the effect that the model "fits" has been confirmed. But whether or not there is a good "fit"[39] is entirely objective. It is a question of the "context of justification."

Though a relatively new conception of scientific theories, the semantic view is not immune from criticism. In fact, many details concerning it have still to be supplied. For one thing, it is not clear on the semantic view how to make sense of the fact that the same scientific theory often has more than one model. For another thing, it does not supply answers to all of the central questions one might want to ask about the nature of scientific theories. It does not, in particular, suggest a clear direction for discussion of questions concerning the reality of scientific entities (some semantic theorists want to say that it implies realism, while others hold just the contrary), nor does it suggest clearly an account of the rationality

[37]Typically, the data are idealized in a number of ways before they ever come to confront a model. See Patrick Suppes, "Models of Data," in *Studies in the Methodology and Foundations of Science*, D. Reidel Publishing Company (1969).

[38]Sometimes it is very difficult to find ways of fitting the model to the data, that is, of finding *actual* physical systems that are isomorphic to some part (the empirical part) of the model.

[39]Utilizing the various marvelously elaborate and subtle techniques of estimating fitness.

of scientific change or of the comparability of theories. It is to the questions about reality and rationality that we now must turn.

5. The Reality of Scientific Entities[40]

An enduring fundamental philosophical question is the question: What is there *really*?[41] After the 16th century, when the postulation of unobservable entities such as atoms to explain the observed phenomena of the natural world became widespread, this question was often replaced by another: To what extent do scientific theories provide us with a correct account of what there really is?

Two answers have been given to the latter question. On the one hand, there are those who claim that scientific theories are a *true* account of the world, and, therefore, the various entities they postulate—atoms, fields, drives, libidos, and so on—are *real*; theoretical objects exist. Those who endorse this claim are called *scientific realists*. On the other hand, there are those who claim that scientific theories are not true accounts of the world; theories are not so much true as *useful*. Since the task of scientific theories is to organize the data of our experience in such a way that predictions about, and eventual control of, the future are possible, theories are simply *instruments*; theoretical objects are by and large convenient fictions. Those who make these claims are called *instrumentalists*.

We begin by setting out three characteristic instrumentalist arguments, then sketch sample realist replies to them, and finally indicate what the crux of the issue is.

[40]What follows incorporates material from Gordon G. Brittan, Jr., "Kant and the Objects of Theory," in B. den Ouden, ed., *New Essays on Kant*, Peter Lang (1987). Defenses of realism include, among many others, Grover Maxwell, "The Ontological Status of Theoretical Entities," in Feigl and Maxwell, eds., *Minnesota Studies in the Philosophy of Science*, University of Minnesota Press (1962), Vol. III, and J.J.C. Smart, *Philosophy and Scientific Realism*, Routledge & Kegan Paul (1964). An anti-realist position is developed by van Fraassen in *The Scientific Image*. Essays on his position can be found in J. Leplin, ed., *Scientific Realism*, University of California Press (1984). Hilary Putnam advances an intermediate "internal realist" position in *Realism and Reason*, Cambridge University Press (1983).

[41]The traditional answers are notorious. A favorite is that of the ancient Greek philosopher Parmenides, whose view seems to imply that there is just one thing in the world and that he isn't it.

Instrumentalist Argument #1: the eliminability of theoretical terms[42]

Assume that one of the principle tasks of science is, in fact, to organize the data of our experience in such a way that predictions about the future course of that experience are possible. What role does the postulation of theoretical entities play? A plausible answer is that it allows us to extend the range of application of our empirical generalizations and to accommodate what would otherwise be exceptions. Many everyday empirical generalizations are limited in the range of their application and suffer from exceptions; for example,

G.1: Wood floats on water, iron sinks in it,

is limited to wood, water, and iron, and suffers exceptions in those cases in which wood (ebony chips, for instance) sinks and iron (perhaps fashioned into spheres) floats. But we can remedy these defects by introducing a theoretical entity, specific gravity, and, by extension, an entity or quantity to which it refers, defined as the quotient of a solid body's weight and volume,

D.1: $s(x) = w(x)/v(x)$.

Then it can be unconditionally asserted that

G.2: A solid body floats on a liquid if its specific gravity is less than that of the liquid.

This generalization, unlike G.1, is unlimited and exceptionless.

Now the difficulty with this interpretation of the function of theoretical entities is that they seem to be eliminable in principle, and therefore we have no reason to assume that they actually exist. That is, we can replace G.2 by

G.3: A solid body floats on a liquid if the quotient of its weight and volume is less than the corresponding quotient for the liquid,

(replacing "specific gravity" by its definition) and make exactly the same predictions; whatever work specific gravity supposedly does can be done without it.

The argument can be generalized. Assume that to make a prediction is to derive an observation statement O_2 from a statement T invoking some theoretical objects and other observation statements O_1, together with the help of certain correspondence rules C. I.e.,

[42]Following Carl Hempel in "The Theoretician's Dilemma," reprinted in Aspects of Scientific Explanation.

O_1
T
C
therefore, O_2.

Clearly, this argument could be replaced by another correlating O_1 and O_2 directly,

O_1
If O_1 then O_2
therefore, O_2,

which yields the same observational consequences but does not make use of statements about theoretical entities. The conclusion would seem to be that we can always get from observations to (predicted) observations without making a theoretical detour. So there would seem to be no reason to suppose that theoretical entities are more than mere instruments, convenient and economical, but in principle eliminable, ways of organizing the data.

This sort of argument has had considerable impact on the methodologies of some scientists. Take the case of psychological explanation. What is given are certain observable aspects of a subject and certain observable stimuli acting on the subject, followed by an observable response. Theoreticians then postulate "intervening variables"—drives, reflex reserves, habit strengths, inhibitions, etc.—to mediate the transition from stimulus to response and in this way to *systematize*, as well as to explain, the observable aspects of the situation. But if the above argument is sound, then a law-like transition from stimulus to response can be made *without* postulating intervening variables. This methodological implication was quickly drawn by behaviorist psychologists such as B.F. Skinner: For purposes of prediction and control, one can proceed directly from stimulus to response without a theoretical detour.[43]

Instrumentalist Argument #2: the incompleteness of theoretical objects

Ordinary objects are such that every assertion about them is either true or false. In a certain sense, then, ordinary sorts of objects are *complete*; the law of bivalence (that every sentence is either true or false)

[43]*Science and Human Behavior*, Macmillan (1953), p. 35.

holds with respect to all sentences ascribing appropriate properties to them. So this book is either red or it is not red, weighs 300 grams or it doesn't weigh 300 grams, etc.

Theoretical objects, however, unlike ordinary sorts of objects, are *incomplete*. The theory in which they are embedded provides no basis for the assertion, for example, that atoms are red or not red, or, in more extreme cases, rules out the possibility that there could be such a basis, for example, that atomic and sub-atomic particles have a jointly and precisely determinable position and velocity. There is no way, even in principle, that such questions could be decided. If we assume, as seems plausible, that an existent object must be complete, then theoretical objects do not, and cannot, exist.

Theoretical objects are often said to be like, even to be, *fictional* entities because they are imaginary. They are like fictional entities because both are essentially incomplete. Many assertions about fictional entities, characters in a novel for instance, are neither true nor false (we can never know, since Willa Cather neither says nor implies an answer, whether Thea Kronberg, the heroine of *Song of the Lark*, had any children). In this respect, and perhaps in others, theories are like novels. They differ, of course, in other ways. Many theoretical entities have properties that are not exhausted by theories about them, spatial-temporal locations for example; these can often be determined by measurement, in which case the theoretical entities in question would have a real history in a way that fictional characters do not. Moreover, theories are supported by evidence in a way in which novels are not.

Instrumentalist Argument #3: the ideality of theoretical entities

Ordinary sorts of objects are "given"; they merely await discovery. Theoretical objects, on the other hand, are mind-dependent; they are invented or postulated, not discovered. They are "free creations of the mind," unlike opaque, resistant ordinary sorts of objects which impose themselves on us.

More specifically, a large subclass of theoretical objects, point-masses for example, are *ideal*. They are treated from the outset as theoretical *constructs*, mere conveniences, with no thought at all that they might exist. Indeed, in such cases as that of a (dimensionless) pointmass, the existence of this sort of theoretical object simply doesn't arise.

These three instrumentalist arguments are arranged in order of increasing strength. The first argument is agnostic: There is no reason to assume that theoretical objects exist, given that all reference to them is in principle eliminable. The second argument is that theoretical objects do not, and cannot, exist because, unlike ordinary sorts of objects, they are incomplete. The third argument is that the question of their existence, at least in exemplary cases, does not even arise; there is simply no issue.

Instrumentalists sometimes add that theirs is the more cautious position. As theories have succeeded one another through time, so have the entities these theories have postulated. Why should one subscribe to the existence of entities postulated by contemporary theories when in all probability some future age will reject them, just as phlogiston, animal magnetism, the ether, impetus, and the funiculus have all been rejected?

These arguments are not immune to criticism. We will begin by considering criticisms of the first and third instrumentalist arguments. Then we will turn to the argument from incompleteness and suggest two different ways in which its implications might be spelled out. Finally, we will consider issues stemming from the apparent instability of the scientific picture of the world.

First, it is a mistake to try to draw a distinction between merely *ideal* and other sorts of theoretical objects. Two celebrated episodes illustrate a recurring pattern in the history of science.[44] One concerns Galileo's postulation of the rectilinear component of the parabolic trajectory. According to Galileo himself, rectilinear motion does not exist; one finds nothing but curved motion in nature. In fact, given the conceptual framework or paradigm against the background of which Galileo's theory of the parabolic trajectory is formulated, not only does rectilinear motion not exist, it *could not* exist.[45] Yet Galileo's treatment, "which he himself considered as an idealization, came to be accepted by his successors as a literally correct treatment; the *true* horizontal component *was* a straight line, and deviation therefrom was due to the action of an external force..."[46]. The second episode is the earlier mentioned postulation of discrete quanta in the analysis of black-body radiation by the 20th century physicist Max Planck.[47] For the curve-fitting Planck, the postulation in question was no more than a calculational device yielding a single formula covering all black-body radiation frequencies. Once again the idealization was self-conscious, for Planck "knew" that in reality

[44]Both are discussed by Dudley Shapere, *Galileo*, University of Chicago Press (1974).

[45]See Alexandre Koyre, *Galilean Studies*, Humanities Press (1978), p. 155.

[46]Shapere, *Galileo*, p. 120.

[47]See M. J. Klein, "Max Planck and the Beginnings of Quantum Theory," *Archive for the History of the Exact Sciences*, 1 (1962), pp. 459ff.

the process had to exhibit continuity, an essential component of every successful physical theory up to that time. Yet following Einstein's later work on the photo-electric effect, discrete quanta came to be accepted as fundamental elements of the true scientific picture of things.

The two episodes show that it does not follow from the fact that certain objects are introduced as "conveniences", or from the fact that no one, at least initially, thinks they exist, indeed even that they could not exist, that the question of their existence does not arise. What these episodes, and others like them, illustrate is that often such objects are eventually and straightforwardly acknowledged to be *real*.

Second, it is a mistake to draw a distinction between theoretical and more ordinary sorts of objects. Note, among other points, that the argument from the eliminability of theoretical objects presupposes a sharp distinction between theoretical and observational objects. But the criticisms made earlier of the classical view of theories jeopardize this distinction given that theoretical terms refer to theoretical objects and observational terms refer to observational objects. Note also that the argument from the eliminability of theoretical objects turns on the claim that the function of such objects is to "systematize" the data in the ways indicated. If this claim were to be rejected, then the argument too would be very much weakened.

Indeed, many realists claim that if scientific explanation and the relation between theories is *correctly* understood, then realism about theoretical entities follows as a kind of corollary. One such realist is Wilfrid Sellars.[48] According to Sellars, theories explain both the phenomena and other theories not by way of "systematization" or derivation but by *identifying* perceptible phenomena with postulated imperceptible entities whose behavior the basic principles of these more fundamental theories describe. According to him, the basic scheme of theoretical explanation is: Ordinary sorts of physical objects "of such and such kinds obey (approximately) such and such...generalizations because they *are* configurations of such and such theoretical entities." If the function of theoretical objects on the instrumentalist view is to "systematize" the data, their function on Sellars' realist view is to *explain* the data. On the former view, it follows that theoretical objects are eliminable in principle. On the latter view, theoretical objects are explanatory *only if* they exist. In other words, to have a good reason to accept a theory is to have a good reason to believe that the entities it postulates

[48] "The Language of Theories," in *Science, Perception and Reality*, Ridgeview Publishing Company (1963, 1991).

in the service of explanation are real. An example will make the point clearer. Suppose an explanation of the fact that gases approximately obey the familiar Boyle-Charles Law (a typical empirical generalization) is requested. An explanation is not primarily a deduction (from the principles of statistical mechanics and the kinetic theory of matter) but an identification of gases with the entities the kinetic theory of matter postulates and whose behavior the principles of statistical mechanics describe. In Sellars' words, "It is because a gas is...a cloud of molecules which are behaving in certain theoretically defined ways, that it obeys the empirical Boyle-Charles Law."[49] A gas becomes hotter when it is compressed because in such a case the molecules of which the gas is composed are forced to travel at greater speeds, etc. If explaining the behavior of some object or type of object involves describing the behavior of the theoretical entities which in some sense compose it, then explanation commits us to the existence of theoretical entities. They cannot be regarded simply as useful fictions; the heart of this kind of explanation is a claim of the form: Such and such physical objects *just are* configurations of such and such theoretical entities.

The picture that emerges from these replies to the first and third of the instrumentalist arguments is that of a continuum of objects, at one end of which are "ideal" objects such as point-masses, at the other end of which are ordinary tables and chairs. To say that it is a *continuum* is to say that there are no sharp breaks in it, no firm distinctions to be made anywhere between "theoretical" and "non-theoretical" objects. But to say this is also to say that we cannot, in principle, distinguish between one and the other as existent and non-existent. The same general types of considerations showing that certain ordinary sorts of physical objects exist can be used to show that some theoretical objects exist as we11.[50]

The second instrumentalist argument, from incompleteness, is more difficult to dismiss. The reason is this. The first and third arguments are characteristically philosophical; they do not appeal directly to empirical theories or experimental data, except by way of illustration. The reply to them, in fact, comes down to this, that if a physical object is possible (its description is not logically inconsistent) its existence or non-existence cannot be demonstrated on merely conceptual grounds. The question is to be resolved by science itself and as a result is at least in part

[49]*Ibid.*, p. 121.

[50]A fascinating account of how the French physical chemist Jean Perrin (1870-1942) established the existence of molecules, conclusively and empirically, is given in Mary Jo Nye's *Molecular Reality*.

empirical. But in the case of the second instrumentalist argument it is a particular, well-founded scientific theory—quantum mechanics—which raises questions about the reality of the entities it postulates.

There is a great temptation to utter the words "quantum mechanics" as a kind of invocation in support of a variety of views in the philosophy of science. Yet there can be no question that the attempt to understand this theory and its implications provides the greatest single challenge to contemporary reflection on the nature and adequacy of the scientific picture of the world.

How, then, does quantum mechanics bear on the argument from the incompleteness of theoretical entities? Suppose we were to argue as follows. According to the molecular model of gases mentioned in the preceding section of this chapter, volumes of gas or ensembles of molecules have a temperature, measurable in familiar ways. Individual molecules, though they have kinetic energy, do not. Therefore, individual molecules are incomplete with respect to the property of temperature. One could always reply to such an argument that statements ascribing a particular temperature to individual molecules of a gas were *false* and in this way restore bivalence: Every meaningful assertion about gas molecules is either true or false. Similarly, statements ascribing colors to atoms would be simply false, and in this way atoms would be completely determined. But an easy out of this kind is precluded in the case of quantum mechanics. For on the dominant or "Copenhagen" interpretation of it advanced by Nils Bohr (1885-1962) and Werner Heisenberg (1901-1976), if a given atomic or subatomic particle *a* has a definite position *M*, then it is neither true nor false that *a* possesses the quantum theoretically incompatible property of occupying a given position *S*. In other words, the particle is *essentially* incomplete; no value can be assigned to its position coordinates.

Briefly put, there are genuine difficulties in the way of maintaining a *realist* position vis-a-vis quantum mechanics, given the "Copenhagen Interpretation." But these difficulties are not insurmountable. The instrumentalist argument turns on two premises: Incomplete objects do not exist and (at least some) theoretical objects are incomplete. Each of these premises can be challenged.

Take the second premise first. There are two difficulties with it. One is that the concept of "completeness" is ambiguous. If it is understood to mean that each real entity possesses always one property out of each (classical) category, then it can be shown[51] that the "completeness" of microphysical objects is compatible with the claim that some statements about them are neither true nor false. The other difficulty is that the Copenhagen Interpretation is not intended to apply to a special class of "theoretical objects." The indeterminacies it postulates hold of objects generally.

There are difficulties as well with the first premise of the instrumentalist argument, that incomplete objects do not exist. One could maintain that it is simply false, since some incomplete objects, photons for example, do exist. One could also maintain that *all* objects are "incomplete" (in our original sense) since bivalence fails with respect to all of them (window glass is neither red nor not red since it is not colored, rainbows are neither here nor there since they are not located, etc.); hence "incompleteness" fails to distinguish between existent and nonexistent objects.

The up-shot is that one can preserve the basic realist strategy of insisting that no sharp distinction can be drawn between "theoretical" and "non-theoretical" objects in the face of the argument from incompleteness, even against the background of the Copenhagen Interpretation of quantum mechanics. For the realist has two clear options. Either maintain that microphysical entities are in the appropriate sense complete, in which case there is no distinction to be made between them and more ordinary sorts of objects (which for purposes of discussion we have assumed are complete) or deny that ordinary sorts of objects are complete, in which case there is once again no distinction to be made between them and the microphysical objects whose incompleteness apparently stems from the Copenhagen Interpretation. Put this the other way around: The thoroughgoing instrumentalist or anti-realist has to show that there is a *difference in kind* between theoretical and more ordinary sorts of physical objects if he or she is to argue that reference to the former can, at least in principle, be eliminated. But "incompleteness" fails to provide an argument on the basis of which such a difference in kind can be demonstrated.

Having said all of this, it should be noted in conclusion that we have, in fact, already granted one cornerstone of the instrumentalist's position. We have maintained that whether a property is observational or not, that whether an object or type of object exists or not, can only be settled by scientific inquiry. Which is to say that there is no independent, extra-

[51]See Karel Lambert, "Logical Truth and Metaphysics," in K. Lambert, ed., *The Logical Way of Doing Things*, Yale University Press (1969).

scientific criterion on the basis of which one can say that scientific entities are *really* real. The question of whether they are *really* real makes little sense from this point of view. Or, as some recent philosophers of science have preferred to put it, our realism about scientific entities is *internal* to science itself.[52]

6. The Rationality of Theory Choice

A final question suggested by the discussion in this chapter is the question whether there are rational grounds for choosing one theory over another, or whether, as historicists sometimes imply,[53] theory choice is more a matter of persuasion and political power than of logic and experiment?

There are two important aspects of this question. One concerns the rationality of scientific *practice*. Are there certain norms in terms of which it can be justified? The other aspect concerns the rationality of scientific change. Is it possible to make out some concept of "progress" that fits the history of science? These sub-questions can be linked. One might claim, for example, that scientific change (one theory replacing another) is rational to the extent that it results from the rational practice of science. We begin, however, by discussing these questions separately.

There are two basic types of account of the rationality of scientific practice, both to be understood via the familiar distinction between *means* and *ends*. One account *emphasizes* the relativity of means. Given that certain ends are chosen or determined in advance, then some means are rational with respect to attaining those ends, others are not. Thus Carl Hempel writes:

> A mode of procedure or a rule calling for that procedure surely can only be rational or irrational relative to the goals the procedure is meant to attain. In so far as a methodological theory does propose rules or norms, these norms have to be regarded as instrumental norms; their appropriateness must be judged by reference to the objectives of the inquiry to which they pertain or, more ambitiously, by reference to the goals of pure scientific research in general.[54]

[52]See Hilary Putnam, *Realism and Reason*.

[53]See Paul Feyerabend, *Against Method*.

[54]"Scientific Rationality: Analytic vs. Pragmatic Perspectives," in T. G. Geraets, ed., *Rationality Today*, University of Ottawa Press (1979), p. 51.

Three features of this account should be noted. First, it is completely general; there are no respects in which scientific practice is particularly or pre-eminently rational *per se*. *Any* activity is rational in so far as it promotes certain desired ends. If Jack wants to go to medical school, then it is rational for him to build up his G.P.A., take a fair number of biology courses, learn to get along on less sleep, and irrational for him to do otherwise. This is what is meant by "rationality"; determine the ends or goals, then it can be decided which courses of action or practices are rational. Second, it follows as a corollary of this point that if only means are rational, then the ends themselves are neither rational nor irrational. Third, if the ends or goals of scientific activity have changed over time, then so too have the means. In fact, the ends or goals do seem to have changed over time. Some scientists (including Newton) have wanted to show the hand of the Creator in the world, others to gain control over nature, still others to satisfy intellectual curiosity. Hempel thinks that the goal of "pure scientific research in general" is to establish "a sequence of increasingly comprehensive and accurate systems of empirical knowledge."[55] But this very contemporary goal, reflective of Hempel's own classical position in the philosophy of science, has not always been pursued. Galileo's theory, for example, is not nearly so comprehensive as Aristotle's, and perhaps no more "accurate" in the sense that it is more precise with respect to ordinary sorts of observations. He advanced it because it was, at least in his eyes, more explanatory and also because he thought that it was true. It follows on the relativity-of-means account that Galileo and Aristotle must have differed concerning which practices are rational, and indeed they did. Aristotle took it as rational to proceed on the basis of sense experience; Galileo, on the basis of mathematical calculations.

It is this sort of point, of course, that historicists emphasize and many of them thus find the relativity-of-means account congenial. Indeed, Kuhn agrees with Hempel that questions concerning the rationality of scientific practice necessarily come down to questions concerning the choice of means to attain ends.[56] From his point of view, a scientist acts rationally just in case he or she acts in ways believed to promote their ends. In this sense, rationality is inevitably agent-specific and context-dependent, in a word "subjective." Only by placing ourselves in the position of the individual scientist, with respect to his or her goals, can

[55]*Ibid.*

[56]See the symposium involving them both together with Wesley Salmon in the *Journal of Philosophy*, 80 (1983).

we determine what it would be rational to do. There is no independent standard of rationality against which every course of action can be evaluated. Neither Galileo nor Aristotle is "more rational" than the other, and getting along on less sleep is not to be recommended on general terms.

The other account of the rationality of scientific practice *rejects* the relativity of means and *emphasizes* the constancy of certain standards and objectives. This account is typically centered around the claim that scientific practice is rational insofar as it is guided by a particular scientific *method*. Thus Israel Scheffler writes:

> It is method rather than doctrine that defines the community of science, and it is the stability of method in pursuit of the truth that holds the community together throughout doctrinal change.[57]

It follows on this account that to the extent that scientific practice is guided by its characteristic method it is *ipso facto* and pre-eminently rational, and that neither ends nor means in science are agent-specific or context-dependent. "Science" is defined by its method. Those who practice science methodically do so rationally.

What is rational about scientific method? Those who take the "methodological" approach claim that, at bottom, scientific method involves a commitment to evidence, however this "commitment" is to be worked out in detail. Various confirmation theories can be seen as ways of doing so. But then the connection between method and rationality is straightforward. For a course of action, or more especially a belief, is rational to the extent that reasons can be given for it, i.e., it can be justified or defended. And we typically justify our activities and our beliefs by supporting them with evidence. Science, on this account, is simply the most systematic and reliable way of supporting beliefs with evidence. Science commands our respect because scientific claims, as a matter of method, must be well-grounded. And this well-groundedness results, in the final analysis, from the application of logic and the carrying out of experiments.

At first glance the "relativity of means" and the "methodological" accounts of rationality appear to be incompatible. But this is not really the case. For, first, the two accounts have a rather different focus. The "relativity of means" account has primarily to do with scientific *activity*.

[57]*Four Pragmatists*, Humanities Press (1974), p. 75. This line is defended by Harvey Siegel in "What is the Question Concerning the Rationality of Science?", *Philosophy of Science*, 52 (1985), pp. 517-37. We draw on his reconciliation of the two accounts in what follows.

We justify our activities in large part by indicating the ends or goals they are intended to bring about, although we also try to provide evidence that in fact the means chosen will bring about the specific ends. The "methodological" account has primarily to do with *beliefs*. To hold a position is to be in a position to defend it. Ultimately we defend our beliefs, rationally, by appealing to evidence, although there is also a connection between what we believe and how we behave. Second, the two accounts emphasize different ingredients in the concept of rationality. From one point of view, a person acts rationally insofar as his or her behavior is consistent. The "relativity of means" account suggests that this activity must be consistent with the results that one is trying to achieve. From another point of view, a person's belief is rational to the extent that reasons can be provided for it, and the best of these reasons is provided by empirical evidence.

In any case, whatever the differences between them, both accounts stress that paradigmatic scientific practice—belief or activity—is rational. One may feel, however, that this conclusion is unhelpful, even trivial, given the way "rationality" has been characterized. For it is difficult if not impossible to characterize scientific "method" any more precisely than we have done so far and the sense in which we have to this point declared that science is rational casts little light on the question of theory choice. In other words, no usable criteria for choosing between theories are implicit in either account of rationality.

Philosophers and scientists have certainly tried to characterize "scientific method" precisely. But it must be conceded that they have not yet succeeded. There seem to be two problems. One is that the different sciences use different procedures, so different that it is impossible to bring them all under a single non-trivial heading. Reflecting on his own practice and that of his colleagues, P. W. Bridgman once said that "science is what scientists do, and there are as many scientific methods as there are individual scientists...". Although within particular scientific disciplines and specific traditions of what Kuhn calls "normal science" it might be possible to isolate certain standards and routines, there seems to be no one method or set of methods that characterizes science as such or which are embraced by successive paradigms. This is presumably the reason why considered statements of scientific method tend to be very general. As Morris Cohen and Ernest Nagel, for example, say in their classic account:

> If we look at all the sciences not only as they differ from each other but also as each changes and grows in the course of time, we find that the constant and universal feature of science is its

general method, which consists in the persistent search for truth.[58]

As we have already noted, the concept of *truth* in science is somewhat problematic and certainly difficult to extend without qualification across the revolutionary changes that have taken place. It is also the case, of course, that a "persistent search for truth" does not mark *science* off as special, nor would this criterion by itself afford a choice between theories which were equally well confirmed. The other, related problem is that there seem to be clear counterexamples to every reasonably precise methodology that has so far been applied to science in general.[59] We have already suggested that sometimes more general theories (Aristotle's) are succeeded by less general theories (Galileo's), even though according to one well-known methodology it must always be the other way around. Similarly, the special theory of relativity was accepted before it was shown that Newtonian mechanics was a special or limiting case, despite the fact that another well-known methodology insists that one theory can replace another in a given domain only if it has been shown that the latter is a special or limiting case of the former. And so on for all of the other classical rules of theory choice or "acceptability," simplicity, elegance, familiarity, etc. None of them (to the extent that they can be made precise) has invariably been at stake when one theory is chosen in preference to another, not even the requirement that the new theory be better confirmed than the old.

The answer to the question "Is scientific practice rational?" is undoubtedly "yes," but the way in which this answer has been spelled out does not add very much to what has already been said in the chapter on confirmation concerning the relation between hypotheses and the evidence for them. Furthermore, the answer does not suggest any sense in which science is a pre-eminently rational activity or in which we can say that progress has been made. We must turn, then, to the question of progress or the rationality of scientific change. It turns out, in fact, that the answers to this second question cast additional light on the first question having to do with scientific practice.

We will consider two views of the nature of progress in the history of science. Both presuppose what seems undeniable, namely, that there has been some sort of "progress." The philosophical task is to say what sort.

[58]*An Introduction to Logic and Scientific Method*, Harcourt, Brace & World (1934), p. 192.

[59]See Larry Laudan, "Progress or Rationality? The Prospects for Normative Naturalism," *American Philosophical Quarterly*, 24 (1987), pp. 19-31.

Our list is not exhaustive, but between them the two views we consider indicate the main lines of approach.

The first view has already been criticised in some detail. It is the "layer cake" view that theories, at least within a particular discipline or domain of inquiry, are deductively linked, more general theories succeeding less general theories over time. We mention it again in order to point out that this view of scientific change is *uniformly goal directed*, that is, presupposes a *goal* in the direction of which scientific change tends. Hempel was quoted at the beginning of the discussion of rationality to the effect that the *goal* of science is the development of ever more comprehensive and accurate theories. A form of this view of science dominated 19th century thought, with particular insistence that at the end of time all natural phenomena should be brought under one big law or theory, although there was a great deal of controversy concerning whether the "one big law" was the law of universal gravitation, the "law" of natural selection ("survival of the fittest") or the second law of thermodynamics. The intuition continues to surface with the suggestion that some grand "unified field theory" will unite, not simply areas of theoretical interest in physics but all domains of human knowledge. The picture here is linear and cumulative, of course, and the line goes in the direction of a final and complete, fully satisfactory account of the world.[60]

A second view of scientific change is *evolutionary* rather than *uniformly goal directed*. Rather than positing some goal at which all scientific activity aims or some final theory which is being more and more closely approximated with the passage of time, the evolutionary approach sees the history of science as a series of adaptations to environmental pressures. The model here is Darwinian. If there has been "progress," it is only in the sense in which organisms make progress, by becoming better adapted. It is from this point of view that Kuhn concludes his argument in *The Structure of Scientific Revolutions*:

> The process described...as the resolution of revolutions is the selection by conflict within the scientific community of the fittest way to practice future science. The net result of a sequence of such revolutionary selections, separated by periods of normal research, is the wonderfully adapted set of instruments we call modern scientific knowledge. Successive stages in that develop-

[60]More will be said about this picture in the chapter on the limits of scientific explanation, under the heading "explanation and unification."

> mental process are marked by an increase in articulation and specialization. And the entire process may have occurred, as we now suppose biological evolution did, without benefit of a set goal, a permanent fixed scientific truth of which each stage in the development of science is a better exemplar.[61]

Although, as Kuhn emphasizes, the parallel between the growth of scientific knowledge and the course of biological evolution can be pushed too far, there is much at least initially to be said for it. It plausibly suggests that development has been *from* sets of ideas rather than *towards* others. It allows us to understand the special status of science, within our own culture and for other cultures as well, by pointing to the ways in which those who engage in scientific activity are more "adapted." Such adaptation, in turn, can be understood in terms of the technological development which science makes possible and which itself has made possible the present economic and political dominance of scientific cultures. It forces us to look more closely at the various selection pressures which bear on scientific theories (including the "data environments" in which they develop) and in terms of which one theory is "fitter" than another. Finally, the evolutionary view makes the development of science a planetary affair. On the "layer cake" model, pushed to its limits, science is going to have to develop along roughly the same lines everywhere in the galaxies. Or, if there are different lines of development, they will eventually have to converge. Whereas if we construe scientific developments as adaptive responses to environmental pressures, then we can say there is no reason why other creatures in other environments might not have adapted in different ways. We know abundantly from our own planetary experience, in fact, that a variety of species can fill the same ecological niches. Nature does not select *for* particular species or, by way of our analogy, *for* particular theories. On this view, there is not some set of inherited characteristics that all and only acceptable scientific theories are going to share.

The evolutionary account of scientific change is at this stage little more than a metaphor. Still, three more specific points of comparison with the classical uniformly goal directed view can be made out. First, the classical view is "externalist" in the sense that it does not understand scientific change in terms of a particular scientific theory, in the case at hand the

[61]Second edition, pp. 172-73. Kuhn sometimes adds that the charge of relativism often made against his position is somewhat mitigated by the fact that on his own evolutionary account of scientific change such change is unidirectional and irreversible; it has a structure which an unqualified relativist would not accept.

theory of evolution. There are theory-independent standards according to which every theory can be appraised. The evolutionary view, to the contrary, is "internalist"; science comes to understand its own development through the application of one of its fundamental theories. On this latter view, there are no independent standards of theory appraisal. In a sense, the later a theory is in time, the more "evolved" it has to be considered, and only in this sense can it be considered "better" than its predecessors. Second, it would seem to follow from this point that whatever is, on the evolutionary account, is rational. Since there is no one end with respect to which the rationality of various means can be assessed and no one method which defines scientific method, there is no *norm* of rationality. That a scientific development takes place over a reasonably long period of time is enough to indicate that there must have been a reason for its survival, that it had certain adaptive characteristics, and in this respect that it was "rational." A third point of comparison is that the uniformly goal directed view posits a goal, whatever it might be, towards which scientific developments, if they are to be at all successful, must tend. The evolutionary point of view is retrospective; given that a species has developed certain modifications, we ask what the selection pressures might have been and how the modifications could have been brought about. In the process we tell a story, a Darwinian natural history, which makes plausible the emergence of these modifications over time. The corollary for the history of science is simply this: Given that our present theories have certain general characteristics, we ask how this might have come to pass. We tell a story from our present point of view which makes the past intelligible. From the present point of view, it seems to be the case that science as a whole has developed with respect to the selection pressures of logical cogency, explanatory power, and experimental confirmation rather than, for example, available funding and political expediency. That is, we can see how our theories developed primarily as a response to the non-human environment and in the process made *us* more adaptive with respect to it. From this point of view, the history of scientific change has been eminently rational. But, on the evolutionary account, no other point of view is possible.

Chapter five

The Nature of Mathematics

1. Introduction

Perhaps the most prominent feature of the preceding chapters is their shared insistence on the *empirical* character of the sciences. The case for the empirical character of the sciences is often argued by contrasting them with such allegedly nonempirical disciplines as mathematics. It is frequently claimed, for example, that mathematics is not a science because "it has no proper subject matter." Rather, the argument goes, its relation to science is more that of form to content: mathematics is the language of science. Its usefulness lies in providing a) means for the precise expression of empirical hypotheses and b) methods for the difficult job of teasing out their consequences. The claim that mathematics is not genuinely a science is often linked to another. In contrast to statements of empirical science—for example, "The rate of response is a function of previous reinforcements"—strictly mathematical statements—for example, "The set of real numbers is nondenumerable"—are certain; no amount of evidence could ever serve to overturn them. Thus Albert Einstein's celebrated dictum: "As far as the propositions of mathematics refer to reality, they are not certain, and as far as they are certain they do not refer to reality."[1]

This view of the nature of mathematics has wide support among contemporary scientists, mathematicians, and philosophers. Why? For one thing, it allows scientists to avoid appeal to what might seem a ghostly realm of nonempirical, abstract "entities" (for example, numbers, points, and sets). For another, it allows mathematicians to distinguish the work they do (so-called "pure" mathematics) and the applications to which that work is put by others. Finally, it allows empirically minded philosophers to account for the apparent certainty of mathematical statements, and at the same time, to deny that knowledge of nature is possible, independent of experience.

One representative of the view that mathematics has no subject matter is the mathematician-philosopher John Kemeny, who offers a particularly straightforward statement of it in his book, *A Philosopher Looks at Science*.[2] According to Kemeny, "Mathematics is a study of the form of arguments and...the most general branch of knowledge, but...it is completely devoid of subject matter" (p. 21). Moreover, "this irreplaceable language of Science, Mathematics, can never supply anything new" (p. 35). To believe otherwise, he maintains, is to confuse pure with applied mathematics; the latter does indeed have a subject matter that varies with the field of application, but the former, which is our present concern does not.

How does Kemeny support this view? He bases it, in effect, on two major claims.

> 1: Mathematical truths reduce to (are nothing but) logical truths.
> 2: Logical truths are analytic, that is, true solely by virtue of the meanings of the words they contain.

It follows from these two claims that true "mathematical propositions are analytic" (p. 21). But if they are analytic, then mathematics is devoid of subject matter, although not in the sense that its statements are not "about" anything. The mathematical proposition

(i) $365 - 1 = 364$

which Kemeny takes as his example, is "about" numbers (if such there be); but the truth or falsity (the truthvalue) of (i), which is the concern

[1]Herman Weyl, on page 134 of his book *Philosophy of Mathematics and Natural Science*, Princeton University Press (1949), quotes this remark from Einstein's book, *Geometrie and Erfahung*.

[2]From *A Philosopher looks at Science* by J.C. Kemeny, Copyright 1959, by Litton Educational Publishing, Inc., by permission of Van Nostrand Reinhold Co.

of the mathematician, does not depend on what it is "about." The truth or falsity of such propositions "depends upon their form," by which Kemeny means "depends solely on how the words they contain are used." To put it another way, the truthvalue of mathematical propositions is determined not by an inspection of nature but by an inspection of linguistic conventions.

For similar reasons, mathematical propositions are certain. "If we try to imagine...ways of testing the ordinary mathematical propositions, we find in each case that it is entirely irrelevant what the result of observation is, we *never* reject the propositions...Mathematical propositions are analytic a priori. They consist of an analysis of the meanings of words" (p. 18).

To get clear about Kemeny's view, it will be necessary to look more closely at the reduction of mathematical truths. Further, we shall have to examine what might be called "the linguistic theory" of logical truth. But first a better understanding of just which truths are logical truths is needed.

2. Logical Truths

Consider the statement "All men are men." Not only is the statement true, it remains true whatever noun or noun phrase we might substitute for "men" (in both of its occurrences). Even if we replace "men" by a noun that presumably does not refer to anything—say, "unicorn"—the statement still remains true. Statements that have this property we call *logical truths*.

This way of indicating the class of logical truths can be made somewhat more precise. If we divide the words appearing in a statement such as "All men are men" into two sorts, logical and descriptive—"all" and "are" are logical words and "men" is a descriptive word—any statement that remains true under any and all substitutions of descriptive words or phrases for the descriptive words or phrases it contains is a logical truth. Alternatively, we could say that logical truths are true statements that contain only logical words *essentially*.[3] Clearly, "All men are men" is a logical truth.

We need to add three points of clarification. The first is that it is extremely difficult if not impossible to give a precise standard for

[3]This rough criterion of logical truth, originally introduced by Bolzano, has been revived and considerably refined by W. V. Quine. See his *Mathematical Logic*, Harvard University Press (1951).

distinguishing between "logical" or "descriptive" expressions. But for our purposes this does not much matter; we can simply count those expressions that logicians class as "logical" as the logical expressions. In fact, four such expressions—"all," "or," "not," and "is-a-member-of"—suffice for the expression of the entire body of mathematics.[4]

Second, our way of stating the criterion suggests that all logical truths contain both descriptive and logical words. But it is important to realize that there are many logical truths that contain no descriptive words. Here are two examples: "There is something that is identical with itself" and "Everything is identical with itself."[5] If these statements are true at all, they must be logical truths, for only logical words occur in them (and so occur in them essentially). Moreover, the existence of such examples is not an idle curiosity. Statements like "There exists something that is self-identical" will prove to be central later in this chapter.

The third point of clarification is that our criterion of logical truth does not commit us to the view that logical truths are true solely by virtue of the meanings of the words they contain, and for this reason are "certain." Indeed, our way of indicating the class of logical truths is perfectly consistent with the alternative views that logical truths delineate certain very general features of the world or that they express fundamental "laws of thought."

In other words, from the fact that we pick statements out as logical truths on the basis of linguistic characteristics—that is, that only logical words occur essentially in them—it does not follow that their truth or certainty derives solely from linguistic considerations. This conclusion requires a separate argument, one which we shall consider when the linguistic theory of logical truth is taken up. Similarly, the fact that all the theorems of geometry are characterizable in a purely linguistic (or "syntactical") way as the set of statements deducible from a given list of axioms does not imply that the truth of these theorems can be traced to purely linguistic origins. For example, plane geometry is often pictured as developing in the following manner. First, surveyors established the

[4]Actually only three words are needed: "or" and " not" can be eliminated in favor of "neither-nor." The reasons for carrying out this discussion with the four mentioned above are pedagogical. In terms of our example, "are" is paraphrased into the primitive logical vocabulary as follows: "Everything either is a man or is not a man." Logical truths containing only "or" and/or "not" are sometimes called *tautologies*.

[5]"is-identical-with" is a logical expression, which can be defined in terms of "all," "or," "not," and "is-a-member-of." This is noteworthy, for it means that "is-identical-with" can be paraphrased only with the help of "is-a-member-of," a term of set theory. There are reasons, to which we shall turn shortly, for not treating set theory as a branch of "logic."

truth of a vast array of geometrical propositions on the basis of observations. Then Euclid systematized this collection by picking out a certain set of these true propositions as basic (the axioms) and deducing the remainder of the collection (the theorems) from this basic set. The point is that the *truth* of the theorems is not linguistic, though the *theoremhood* of these truths unquestionably is.

3. The Reduction of Mathematics

The first of the two claims on which the view of mathematics we are considering is based is that mathematical truths reduce to(are nothing but) logical truths. Typically, the claim is supported by showing that all mathematical statements are deducible, with the help only of logical rules, from a handful of logical principles. To show this is to show that any ostensibly mathematical statement can be paraphrased in such a way that the only words that occur in it essentially are logical words. Mathematics reduces, in this sense, to logic. The attempt to show this—a program known as *logicism* and associated with the names of Frege, Russell, and Whitehead—is one of the great intellectual adventures of modern times. The logicist program is set forth in greatest detail in Russell and Whitehead's *Principia Mathematica*.[6] Here we must content ourselves with the barest sketch of it.

The first step in the program is to reduce the various branches of mathematics, for example, analysis, algebra, geometry, to arithmetic, on the model of Descartes' reduction of geometry to algebra via analytic geometry. As Kemeny puts it: "It can be shown that all of Mathematics is founded on properties of integers (whole numbers). If you are well acquainted with these, the rest of Mathematics is deducible by purely logical arguments. So in a sense the nature of Mathematics can be identified with that of the theory of integers" (p. 20).

The second step is to show how arithmetic, "the theory of integers," can be reduced to logic. In *Principia*, this is accomplished somewhat as follows: We take it that arithmetic can be developed on the basis of five axioms set down late in the last century by the Italian mathematician Peano; from them all the properties of the integers can be derived by strictly logical reasoning.[7] The five axioms are:

[6]Gottlob Frege, *The Foundations of Arithmetic*, (trans. J. L. Austin) Harper and Row (1953), is an excellent introduction to the logicist program.

[7]Actually, these five axioms suffice for only a fragment of arithmetic, but the point in question is not affected.

A.1: 0 is a number.
A.2: The successor of any number is a number.
A.3: No two numbers have the same successor.
A.4: 0 is not the successor of any number.
A.5: If P is a predicate true of 0, and if whenever P is true of a number n, it is also true of the successor of n, then P is true of every number.

The trick is to define the arithmetical notions that appear in these axioms in logical terms. Once this is done, the reduction of mathematics to logic proceeds without difficulty. Once numbers are construed in logical terms, the basic arithmetical operations (addition and multiplication) can be similarly defined. In this way, mathematical statements turn out to be *provable as theorems in logic.*

The three notions involved are "0", "is a number," and "is the successor of." In fact, the second can be defined in terms of the first and the third. To say that n is a natural number is to say that n is 0 or is the successor of 0 or is the successor of the successor of 0, and so on.[8] We have only to deal with "0" and "is the successor of." "0" can be defined as the set that contains only the set that has no members—that is, the set that contains the null set as its only member. And the successor of any number n is the set of all sets that, when deprived of a member, come to belong to n. Since these defining expressions in turn can be analyzed in terms of "all," "or," "not," and "is-a-member-of," which are logical expressions, the reduction of mathematics to logic is virtually complete. To quote Kemeny once more, Russell and Whitehead "show that the mathematical concepts used by Peano can be defined in terms of logical words and that all their properties can be demonstrated by pure logic. Thus Mathematics is shown to be no more than highly developed Logic" (p. 21).[9]

[8]One of Frege's principal contributions was to analyse out the "etc." On his version as simplified by Russell, n is a natural number just in case n is a member of every set X such that 0 is a member of X and all successors of member of X are members of X. See W. V. Quine, *Set Theory and Its Logic*, Harvard University Press, #12, for a discussion of Frege's definition and the reduction of arithmetic.

[9]Kemeny goes on to add the following crucial proviso: "In this process two new logical principles turn up, the axioms of infinity and choice, whose somewhat controversial nature need not concern us here. Let it suffice that if we recognize these two as legitimate logical principles—as most logicians do—then all of Mathematics follows and becomes just advanced Logic."

4. The Linguistic Theory of Truth

The second claim on which Kemeny's argument turns is that logical truths are analytic, that is true by virtue of the meanings of the words they contain. As we have already mentioned, our characterization of the class of logical truths does not commit us to any special doctrine about where their truth comes from, nor does it enable us to infer that logical truths are "certain." The linguistic theory of logical truth is such a special doctrine and it does allow us to draw the intended conclusions. There are other answers—for example that logical truths are certain because they describe the most general and pervasive traits of reality—but, as we have also had occasion to mention, an answer of this kind does not recommend itself to the empirically-minded philosopher.[10]

Proponents of the linguistic theory say that we are to regard logical truths as a limiting case in the class of true statements. Many true statements have both a factual and a linguistic component. Thus, the truth of "The grass is green" depends both on what I mean by "grass," "green," and so on, and on whether or not the grass is in fact green. The truth of the statement "The grass is green or it is not green," however, seems to depend solely on how the words that it contains are used; that is, the truth of the statement in question depends solely on the meanings of the logical words "is," "or," and "not," and on the understanding that "green," in both of its occurrences, has the same meaning (whatever it may be). In the same way, logical truths generally owe their truth to linguistic, not factual, sources; their truth is determined solely by the ways in which the words they contain are used. For example, if "The grass is green or it is not green" were false, it would be because the words it contains had meanings different than the ones we normally associate with them, not because the world has been misdescribed. This theory about logical truth is sometimes loosely stated by saying that logical (and ultimately mathematical) truths are "true by definition." Since they are "true by definition," moreover, such truths cannot be overturned by empirical evidence; they are certain.

The linguistic theory is also sometimes stated as the theory that logical truths are "true by convention." In the sense that all definitions are conventional, of course, this variant of the theory is closely tied to the one discussed in the preceding paragraph. What it comes to is this: on our criterion, logical truths are those true statements that contain only

[10]Some of what follows derives from W. V. Quine's essay, "Carnap and Logical Truth" which is in *The Ways of Paradox*, Random House (1966).

logical words essentially; but, the argument goes, the meaning of these logical words "or," "not," and so on, is fixed conventionally. For example, logicians decide that statements compounded with "or" shall be true just in case at least one of the components is true. It is because the meanings of the logical words have been fixed in certain ways that the sentences in which they alone occur essentially are true. Conversely, if we had adopted a different set of conventions concerning their use, the class of truths in which they alone figure essentially would not be the same. But, once again, if logical truths are "true by convention," then it seems to follow that they must be certain, that their truth is independent of how matters stand in the world.

Taken together with the first claim, that mathematical truths reduce to logical truths, the linguistic theory extends to mathematical truths as well. Thus Kemeny: a (true) mathematical proposition is "true because of the meanings of the terms; it is a true analytic proposition" (pp. 21-22). We know that, for example, "365 – 1 = 364" is true when we know what "365," "–," "1," "=," and "364" mean. But knowing what these mathematical expressions mean comes, if mathematics reduces to logic, to knowing what the logical expressions in terms of which they can be defined mean. But we do know what these expressions mean; their meanings have been determined in accordance with certain usages. Similarly, the truth of the theorems of Euclidean geometry, for example, the theorem "A straight line is the shortest distance between two points," is determined by the fact that "point," "line," "plane," and so on have the meanings they do, and not, for example, by the (alleged) fact that the physical world is Euclidean in structure. This, to repeat a point made earlier, would be to confuse pure with applied mathematics. Statements of applied mathematics are not true solely by virtue of the meanings of the constituent words; that is, they are not analytic, but neither are they "certain."

A final point. The above claims are not unconnected. How logical truth is characterized and how the reduction at stake is to be appraised both turn in large part on what we understand by "logic." Conclusions will vary, depending on the latitude we allow ourselves regarding this term.

5. Logic and Mathematics

Notice that the reduction of mathematics, by way of defining "0" and "successor" in logical terms, involves the notion of a *set*.[11] If mathematics is reducible to logic, then "logic" must be understood to include set theory. There are reasons, however, for distinguishing between logic in a narrower sense and set theory, and ultimately for distinguishing between mathematics and logic.

For example, if we were to look more carefully at Russell and Whitehead's program in *Principia*, we would see that it depended essentially on the use of two axioms—the axiom of infinity and the axiom of choice—that are extremely difficult to construe as principles whose truth or falsity depends solely on the meanings of the constituent words. That there exists an infinite number of individuals, which is what the axiom of infinity asserts, seems to be true neither by virtue of the meanings of the words "exists" and "infinite" nor by virtue solely of the meanings of the logical words involved. The axiom is true, if at all, just in case it is a fact that there exist an infinite number of individuals. Similarly, the axiom of choice requires more than an imaginative stretch to be regarded by one as a principle of logic.[12] This point is worth developing. Both of these controversial axioms make existence claims. In fact, set theory generally contains a variety of assertions concerning the existence of certain sorts of things. Consider the statement that there exists a null set, a set with no members. This statement (which contains no descriptive words, by the way) certainly appears to be about something—namely, the null set. So those who contend that mathematics is not really about anything, because the truth of the statements with which it is concerned stems from the arbitrarily assigned meanings of the words making them up, must argue that the reliance of the truth of such statements on what they are about is merely apparent; a way of speaking at best. But suppose there were no objects at all or at least no sets. Then the statements that *there exists* a null set could not be true. Whether there is anything, whether something exists depends on the facts and is certainly not solely dependent on how we use words. That neutrinos, or

[11]The notion of a set and the elementary principles of set theory are now part of most high-school mathematics programs. A set is a collection of things: it has members. In other words, set theory involves the expression "is-a-member-of." This expression cannot be analyzed in terms of the other three basic logical words "all," "or," and "not." It follows that set theory cannot be reduced to a logic that talks only about individuals.

[12]The axiom of choice asserts that there is a way of picking a single member from each of a list of disjoint sets such that the selected members comprise a new set distinct from all others in the original list. The truth of this axiom, if it is true, requires that there *be* such a way (that is, such a function); it is not enough that the words involved in its expression have the meanings they do.

habits, or even God (*pace* St. Anselm) exist(s) is not decidable by appeal to only the meanings of the words "neutrino," "habit," "God," and "exist(s)." However these statements are to be established, their truth or falsity does not come from the use of words alone. The claim that the truth of, for example, "There exists a null set" only *appears* to depend on what it is about is a claim that itself rests on an illusion.

An objection that thus arises immediately to Kemeny's view of the nature of mathematics is simply that many mathematical truths—indeed, many ostensibly "logical" truths (witness "There exists something that is self-identical")—are not analytic. Thus, insofar as these statements depend upon the "facts" for their truth, mathematics does have a subject matter and what it is about—sets, numbers, or whatever—is an important area of investigation. This objection is directed more against a certain theory of logical truth—we have called it the linguistic theory—than against the claim that mathematics reduces to logic. Yet the above remarks bear on this last claim also.

According to Leibniz, the seventeenth century mathematician and philosopher, a logical truth is one that is true in all possible "worlds." The notion of a possible world is a difficult one; contemporary philosophers of logic tend to visualize a possible world as a set of objects with their properties. But then one possible world is the world that contains nothing. In recent years, logicians of a philosophical bent have constructed logical systems, all of whose theorems hold in all possible worlds *including the empty one*.[13] Now this reformulation of logic in accordance with Leibniz' characterization of logical truth as truth in *all* possible worlds has the effect of eliminating statements like "There exists something that is self-identical" and, indeed, any statement that begins with "There exists a..." as truths of logic. The statement in question would only be true in a "world" having at least one member, but would be false in the empty world. Since the Quine-Bolzano standard of logical truth does not exclude such statements from the class of logical truths, it may reasonably be claimed not to be an equivalent formulation of Leibniz' characterization so long as possible worlds are construed as *sets* of objects.[14] There is, however, a deeper point. Logic reformulated in the

[13]See Karel Lambert's paper, "Free Logic and the Concept of Existence," in the *Notre Dame Journal of Formal Logic*, Vol. VIII (April, 1967), pp. 135-41, for an example of such a system.
[14]See the paper "Logical Truth Revisited" by P. Hinman, J. Kim, and S. Stich in the *Journal of Philosophy*, LXV, (September, 1968), 495-500, for a good discussion of the Quine-Bolzano standard. See also D. Berlinski and D. Gallin, "Quine's Definition of Logical Truth," *Nous*, III, No. 2, (1969), 111-28.

above way contains no existence claims. Since set theory abounds in them, it is not the case that mathematics reduces to logic, because the alleged reduction takes place, so the argument goes, via set theory.[15]

Whether logic contains existence claims or not, there is another argument often advanced to show that mathematical truth does not reduce to logical truth, an argument based on a celebrated discovery by the mathematical logician Kurt Gödel.[16] According to Gödel's "incompleteness theorem," there are mathematical truths that are not provable within the resources of a formal system designed to yield such proofs. In short, mathematical truth does not coincide with proof in a formal system. (It would be a mistake, of course, to think this means that some mathematical truths are *absolutely* unprovable.) The point is that logic can be formalized—that is, it can be organized in such a way that its theorems (provable formulas) can be got from certain explicit axioms by certain syntactical rules. Further, it can be shown that the class of theorems coincides with the class of logical truths, a result also first proved by Gödel. Accordingly, if mathematical truth reduced (by derivation) to logical truth, mathematical truth would coincide with proof in a formal system, namely, formalized logic. But this is just what Gödel's "incompleteness theorem" precludes. Hence, mathematics cannot reduce to logic so long, once again, as logic is thought of as not including set theory. This objection presents us with a dilemma: Either logic does not include set theory, or the class of logical truths (understood to include the truths of set theory) does not coincide with provability in a formal system. If one is prepared to accept the unprovability of a whole class of logical truths, then one need not boggle, on that account, at the inclusion of set theory in logic.

6. Truth by Convention

Kemeny's view about the conventional character of mathematical truth is also open to objection. Specifically, it might be urged, he seems to misplace what really is conventional and arbitrary in logical and mathematical statements. Consider once again the statement that there

[15]Earlier in this chapter, Frege was mentioned as one of the founders of Logicism, the position now being criticized. Apparently, in his later years, Frege rejected Logicism on grounds similar to those presented in this, and in the preceding paragraphs of this section. See his "Neuer Versuch der Arithmetik" in the *Nachgelassene Schriften*, Harmes, Kambertel, and Kaulbach, eds., Felix Meiner Verlag (1969).

[16]There is a clear and useful account of Gödel's discovery in Ernest Nagel and James R. Newman, *Gödel's Proof*, New York University Press (1960).

exists a null set. In most versions of set theory, this statement, if not an axiom itself, is deducible from the axiom that says: for every property there exists a set consisting of all and only those objects that have that property.[17] So the statement that there exists an empty set has a subject matter provided the axiom from which it is derived has a subject matter. It is at this point that the misplacement enters. It is *not* the truth of the theorem that is arbitrary or conventional, but rather the particular truth that we choose to be an axiom. To put it another way, *axiomhood* is conventional, *truth* is not. So much is hidden by the expression "true by convention": The expression might mean *either* that the truth of a statement is conventional *or* that it is a particular true statement chosen to be an axiom that is conventional. But even a cursory examination of the history of mathematics suffices to show that what is arbitrary in the behavior of mathematicians has to do with which truth (or truths) in a batch of truths is taken as a sufficient basis for producing the rest of the statements in a given branch of mathematics. Euclid's geometry is a prime example. This objection to the "truth by convention" thesis is suggested by W. V. Quine.[18]

Another objection is that the linguistic theory of logical truth is vitiated by a subtle ambiguity. Recall that the theory asserts that the truth of a truth of logic is *determined* solely by the meanings of the constituent words. Equivalent statements of the theory may be obtained by replacing "determined" by "depends on" or "in virtue of" (and rephrasing the rest of the statement, if need be, to make grammatical sense). The point is that the linguistic theory seems to be unavoidable when one examines the methods logicians use in determining logical truth. For example, let us look again at propositional logic. A classical method used to determine (propositional) logical truth is the so-called tabular method. Very roughly, it can be described as a method for computing the truth of compound statements on the basis of the truth values of their simple components. Thus suppose that one takes the statement "John is tall" as the only simple (or *atomic*) statement in a language where the means of constructing compounds are the conjunctions "not" and "or." (An atomic statement is one which contains neither "not" nor "or.") Now we may lay

[17]It should be remarked that unless restricted this axiom leads straight to Russell's Paradox. (Take the defining property to be: is not a member of itself. Then the set of all sets that are not members of themselves is a member of itself just in case it is not a member of itself, and not a member of itself just in case it is a member of itself). But the needed restrictions do not bear on the issue at hand.

[18]W. V. Quine, "Carnap and Logical Truth," in *The Ways of Paradox.*

down "conventions" for computing the truth and eventually the logical truth of compounds like "John is not tall" or "John is tall or John is not tall," and so on. The "conventions" are essentially threefold:[19]

> 1. Assign every simple statement either the value "true" or the value "false" (on the assumption that every statement is either true or false).
> 2. "Define" the logical conjunctions "or" and "not" in such a way that one can tell the truthvalue of a statement containing these conjunctions (for example, we can "define" a statement of the form "___is not..." as true when "___ is..." is false, and vice versa; similarly, a statement of the form "___ or..." is true just in case at least one of "___" or "..." is true, otherwise false.
> 3. Compute the truthvalues of the compounds on the basis of 1. and 2. (for example, if we know the truthvalue of "John is tall"—say it is true—then "John is not tall" must be false, given our definition of "___is not...").

Now certain statements will turn out true by this method no matter what the truthvalues of their component atomic statements. The statement "John is tall or he is not tall" is such a statement, because if "John is tall" is true, the compound is true, and if "John is tall" is false, the compound is still true.

Note that the statement in question is also logically true by either the Quine-Bolzano or the Leibniz ("possible worlds") standard. Accordingly, the tabular method enables us to pick out a large class of logical truths. Now as we mentioned above, it seems perfectly appropriate to say that statements like "John is tall or he is not tall" have their truth determined solely by appeal to the meanings of their constituent (logical) words—in this case by appeal to the meanings of the words "not" and "or." The linguistic theory interprets this perfectly appropriate claim as a statement about the *source* of logical truth. But it is just as plausible to say that what the tabular method does is to "determine" logical truths merely in the sense of picking them out, and that the truth of logical truths does not come from the linguistic "conventions" (rules and definitions) used to evaluate compound statements. The same may be said for the linguistic theory of logical truth expressed in terms of "depends on" or "in virtue of." In other words, the "conventions" associated with methods for

[19]The present conventions are thus conventions that allow us to evaluate the truth of any statement, simple or compound, in which the logical words may occur either essentially or inessentially.

ascertaining logical truth can be thought of as *devices for identifying* (picking out) logical truths, in the way that fingerprints are used to identify men, and not as the *sources* of the truth of statements such as "John is tall or he is not tall." The objection, then, is that the linguistic theory of logical truth is at best unproved, and its force is considerably diminished when the ambiguity of "determined" and its synonyms are noted.

7. Mathematics and Science

The argument we have been considering has two main premises—one about the assimilation of mathematics to logic and the other about the linguistic theory of logical truth. Neither, as we have seen, is uncontroversial. But that is not the end of the matter; for it still might be felt that, however unconvincing the argument, the intuitive contrast it was intended to support—between scientific and mathematical statements—is real. Mathematical statements differ sharply from scientific statements in that only the former are *certain* and that *empirical evidence is irrelevant* to their truth or falsity. In this section we shall discuss very briefly alternative accounts that many have held to support the contrast successfully.

The first of these touches on a point that we have already made. We suggested that, in one sense, mathematics does have a subject matter and is not to be distinguished in this respect from science, but that its subject matter—sets for one thing—seems to be different. To put it in an almost question-begging way, mathematics has a nonempirical subject matter, whereas the subject matter that is the concern of scientists consists of observable, spatially and temporally locatable objects and events. Mathematics seems to be about objects that are unobservable and that have no spatio-temporal location. This is what is usually meant, for instance, by the assertion that sets are *abstract* objects.

Of course, if anyone were to show that mathematics could dispense with abstract objects—that in the case at hand one could eliminate sets in favor of spatio-temporal objects then the alleged contrast with science would collapse.[20] But there are more immediately available reasons for

[20]Interestingly, Russell made an attempt to get along without sets in *Principia Mathematica*, on the basis of a so-called "no class" theory that appeared to be committed to the existence of individual objects only. He failed, however, to do away with abstract objects. His elimination succeeds only at the expense of introducing *attributes*, which is just another variety of abstract object. See Charles Chihara, *Ontology and the Vicious-Circle Principle*, Cornell University Press (1973).

rejecting the present contrast between mathematics and empirical science. On the one hand, it is at least very controversial whether observable, spatio-temporal objects and events exhaust the subject matter of science (properly so-called). Many of the objects (for example, submicroscopic particles) that science studies are not in any straightforward sense "observable," and in the more theoretical reaches of science many qualifications have to be made about "spatio-temporal locatability." On the other hand, it has been claimed[21] that, while they are perhaps not spatio-temporal, mathematical objects are in some sense "observable." Moreover, it seems difficult if not impossible to contrast these senses of "observable" without begging the question.[22]

There is a related way of drawing a line between science and mathematics—not in terms of their subject matter, but in terms of the *methods* used to demonstrate the truth and falsity of statements. Mathematical statements are shown to be true or false only by analytical procedures, for example, like deduction; empirical statements in contrast require observation to demonstrate their truth or falsity.

A moments's reflection, however, should make it clear that distinguishing between mathematical and scientific statements on the basis of the methods used to establish them will not work. Many of the "higher order" statements in physics, for example, are not directly testable by observation, nor are immediate consequences of these statements. Often such statements are justified by the simple fact that they are consequences of other, already accepted statements. Indeed, we can go so far as to say that it is doubtful whether a "higher order" statement would be accepted purely on the basis of empirical evidence. But if the acceptability of many statements of science is simply a consequence of their implying or being implied by other already accepted statements of a theory, then there is no contrast with mathematics; for whether a statement implies or is implied by another is determined by strictly analytical procedures.

There remain those who protest that, although it is perhaps only very indirect, there is a connection between empirical matters of fact and scientific statements that does not obtain in the case of mathematical statements. Somewhere in the hierarchy of statements that make up a

[21]Most notably by Kurt Gödel. See his paper "Cantor's Continuum Problem," first published in 1947 and reprinted in Benaceraff and Putnam, eds., *Philosophy of Mathematics: Selected Readings*, Prentice-Hall (1964).

[22]Gödel puts it this way: "despite their remoteness from sense experience, we do have something like a perception also of the objects of set theory, as is seen from the fact that the axioms force themselves upon us as being true. I don't see any reason why we should have less confidence in this kind of perception, i.e., in mathematical intuition, than in sense perception..."*Ibid*, p. 271.

scientific theory there is contact with the world, so that at least some of the statements have observational consequences; but the same is not true of any system of pure (that is, not applied) mathematical statements. This claim has to do more with the statements of mathematics and science considered collectively than individually, and in this respect it differs from the arguments already considered. There seem to be at least two way in which it can be interpreted.

One way proceeds in terms of a notion of "factual enrichment." Consider the following three statements:

1. John is tall.
2. Henry is fat.
3. The rat ran right.

We label these *atomic* (or *simple*) statements because they are not compounded of other statements using the logical operators among which are "or," "not," and "all," and we can also take them to be *observation* statements, in the sense that their truth or falsity can more or less be determined by observation. Now an interesting fact about logic can be noted: It is not possible to enlarge this list of the three atomic statements by means of the logical words alone and the rules governing their use. Logic can tell us how to compound these statements (in truth-preserving ways), but it cannot add other atomic statements to our basic stock. This is what one means when one says that logic is merely a set of rules for transforming statements into other statements, and for this reason is not factually enriching.

Furthermore, even if we add mathematics to logic, we cannot get any additional simple observation statements than the three we began with. The addition of mathematics does not increase our ability to get new factual information of this type. Thus mathematics too is factually unenriching. The proof of this claim is not difficult.[23] Briefly, it can be shown a) that three axioms of set theory—the so-called axioms of comprehension (suitably restricted), extensionality, and choice—added to the principles of logic suffice for the development of mathematics, and b) that the addition of these three axioms to logic does not permit us to derive any more atomic observation statements than we could before they were added. In view of their inability to increase our factual knowledge in this sense, mathematical statements may be said to be

[23]Those interested can find it in an article by Hilary Putnam, "Mathematics and the Existence of Abstract Entities," *Philosophical Studies*, VII (1956), pp. 81-88. For Putnam's second thoughts on the matter, see his *Philosophy of Logic*, Harper Torchbooks (1971).

contrasted with at least some of the more or less "observational" statements that make up a given scientific theory.

From this point of view, and not from the questionable thesis that mathematics has no subject matter, there may be something to the claim that mathematics is the language of science. It provides the means whereby facts can be expressed and the relations between them made clear. If mathematics alone cannot lead us to new, simple observation statements, it can greatly enrich the ways in which these statements can be expressed and combined.

At the same time, this argument does not appear to take us very far toward supporting the contrast between mathematical and scientific statements in terms of certainty. It gives us no reason to suppose that mathematical statements are any more or less certain than scientific statements.

The second way to interpret the claim that mathematical statements do not stand in the same relation to experience as empirical scientific statements is that the former cannot be refuted by the facts. They are true, if at all, come what may; empirical scientific statements, on the other hand, are always subject to revision in the face of recalcitrant experience in the laboratory, and so are not certain or necessary. At this point, we make contact again with Kemeny's view of the nature of mathematics. Mathematical statements for Kemeny—and for those who share his position—are true or false independent of how matters stand in the world; they are *analytic*. Empirical scientific statements, on the contrary, are *synthetic*: They are not true because of their form, but because of their content, that is, because of the way in which they are confirmed by observation, however indirectly.

This view is very widely accepted as true, although it should be noted at once that as it stands it is not very precise. We have already had something to say about the lack of clarity in "observation." Switching to an alternative vocabulary and saying that mathematical, in contrast to scientific, statements are *necessary* is not very helpful either, for in the attempt to make *this* expression clear, we are quickly led back to talk about "confirmed" and "observation" (for example, a statement is necessary just in case there is no possibility of its being disconfirmed by observation, and so on).

There are powerful objections to the view that there is a sharp distinction between mathematical and scientific statements on the basis that the former are analytic and the latter synthetic. Heretical as it might sound, a case can be made for saying that mathematical statements too are "always subject to revision in the face of recalcitrant experience in the laboratory."

This case can best be appreciated by way of examples. We shall focus on two. Consider the statement that momentum is proportional to velocity as one paradigm of an analytic statement.[24] "Momentum" is here simply *defined* as "mass times velocity"—that is what "momentum" *means*. Now suppose this statement to be part of a scientific theory (for example, classical physics) with which certain experimental findings (for example, the Michelson-Morley experiment) might seem to conflict.[25] Since these findings do not conflict with any one statement of the theory in particular, there are a variety of ways in which the theory might be revised in order to accommodate them. One such revision that serves to align theory with data amounts to amending the statement that momentum is proportional to velocity. We might say (as a matter of definition, or, better yet, redefinition) that momentum is proportional to velocity divided by one minus the ratio of the velocity to the speed of light. This allows one to accommodate the Michelson-Morley experiment, among others, into a physics that is in many other respects "classical." But if an "analytic statement," as the one in question ostensibly is, can be revised in this way when the theory in which it belongs is confronted by recalcitrant experience, there seems to be no reason for contrasting it with other, ostensibly "synthetic" statements, which the theory also contains.

An even more striking case concerns contemporary developments in physics, in particular quantum mechanics. This time our paradigmatic analytic statement is the law of excluded middle, "*p* or not *p*." This is a fundamental logical principle, if there are any. It meets every criterion one might want to propose for an analytic statement—logical truth, necessary proposition, and so on. But according to Heisenberg's "Uncertainty Principle," a simultaneous determination of a particle's

[24]The example is W. V. Quine's and what follows draws heavily on his article "Necessary Truth," first published in 1963 and reprinted in *The Ways of Paradox*. See also H. Putnam, "The Analytic and the Synthetic," in *Minnesota Studies in the Philosophy of Science*, Vol. III, eds. H. Feigl and G. Maxwell, University of Minnesota Press (1962), pp. 358-97.

[25]The Michelson-Morley experiment was designed to determine whether the earth moved with respect to an ether in which it was hypothesized to be located. By the end of the 19th century, the ether hypothesis had come to be incorporated into classical physics, and the failure of the experiment to find a relative velocity was taken as evidence against the classical theory and in favor of Einstein's special theory of relativity which had abandoned the ether hypothesis. The experiment is one of the most famous in the history of science, and its correct interpretation continues to be controversial. The original paper is Michelson and Morley, "On the Relative Motion of the Earth and the Luminiferous Ether," *American Journal of Science*, 34 (1887), pp. 333-45.

momentum and position is impossible. It is, for example, not possible (on that principle) to maintain that an object has, at a given moment, such and such momentum and position or that it does not. As a result, certain philosophers[26] have recommended abandoning the law of excluded middle and choosing instead a logic in which it does not figure as a logical truth, which is, once again, to revise even one's logic in the face of recalcitrant experience.[27]

These two cases suggest that although at any given time, within any particular scientific theory, certain statements are not held liable to revision, there is no reason why they should not be. Thus, if certain predictions in a discipline such as mechanics fail, one does not typically embark on a revision of the differential calculus, though the calculus is part of that theory. But there are circumstances in which one might conceivably do so. Statements that are, relative to certain contexts, immune to revision in the face of the facts might be called "analytic" or "certain." The important point is that whether such statements are construed as "analytic" depends on the context. A statement can be taken as definitional in one interpretation of theory, as a natural law in another interpretation, and so on. Newton's second law—$F = ma$ (force equals mass times acceleration)—has been variously interpreted in just these ways. The truth or falsity of the statement is not simply determined either by its form or by its content, but rather by the role it plays in those theories in which it figures. Thus one can, relative to certain theories and contexts (for example, when the theory is being tested), make a distinction between "analytic" and "synthetic" statements. The

[26]Notably Hans Reichenbach in his book *Philosophical Foundations of Quantum Mechanics*, University of California Press (1944). Reichenbach developed a 3-valued logic as the logic of quantum mechanics. Other philosophers, those of the so-called "intuitionist" school, have recommended abandonment of the same law in mathematics, on the basis of quite different considerations.

[27]This is not the only, or even the most widely shared, response. For an introduction to quantum logic (the logic of quantum mechanics), see Max Jammer, *The Philosophy of Quantum Mechanics*, Wiley (1974), Chapter 8. H. Putnam's radical interpretation of quantum logic is presented in "Is Logic Empirical?" The mathematical physicists J. von Neumann and G. Birkhoff published their classic on the topic "The Logic of Quantum Mechanics" in 1934. Both essays may be found reprinted in C.A. Hooker, ed., *The Logico-Algebraic Approach to Quantum Mechanics*, vol. 8, Reidel (1975). Also reprinted there is Bas van Fraassen's illuminating essay "The Labyrinth of Quantum Logic" which compares various approaches to the topic. For an attempt to preserve Reichenbach's view within the framework of standard logic, see Karel Lambert, "Logical Truth and Microphysics," in K. Lambert, ed. *The Logical Way of Doing Things*, Yale University Press (1969).

"analytic" statements will be those held *constant*:[28] recalcitrant experience will have no bearing on their truth or falsity. In most cases the statements held constant will be the logical and mathematical ones.[29] But there are cases like those sketched, in which even logical and mathematical principles may be subject to revision. Mathematical statements do not typically stand in the same close relationships to experience that empirical statements do, and they are to that extent "more necessary," less subject to revision—but the difference seems to be one of degree only. Any sharp boundary between mathematics and science is, perhaps, no more than terminological.

[28]An interesting characterization of "analytic" in this spirit may be found in Bas van Fraassen's paper, "Meaning Relations Among Predicates" in *Nous*, I (1967), pp. 161-81.
[29]One reason we are so reluctant to revise them is because of their very great generality.

Chapter six

The Limits of Scientific Explanation

1. Introduction

It has long been a matter of controversy whether there are limits to what can be explained by science. The issue is many faceted. So the first task in this final chapter is to stake out the exact question to be discussed.

We shall not be concerned with the question whether science can explain anything at all. So we shall not address the argument of the extreme skeptic who complains that since the reliability of the materials appealed to in scientific explanations—theories, laws, hypotheses, etc.—rests on inductive inferences from data, a procedure which has no rational justification, science can explain nothing at all. Nor shall we be concerned with the question whether there is anything at all that science cannot explain. So we shall not examine the argument of the extreme optimist who urges that because the development of science—since the 16th century at least—has been marked by the extension of scientific explanations to ever new and broader kinds of phenomena, that there are in principle no limits to the scientific explanation of *every* phenomenon.

We shall be concerned instead with the question *whether there are certain kinds of phenomena inside the agreed scope of science that cannot be explained scientifically.* The worth of any argument purporting to establish such an immensely important conclusion will depend, of course, in large measure on the underlying theory of scientific explanation. The necessary preamble was undertaken in Chapter II and will be used in the ensuing discussion.

We shall confine ourselves to two well known (and widely expressed) arguments on the limits of scientific explanation. They will suffice to give the reader the flavor of the debate and, hopefully, to provoke further investigation of the profound and sometimes disturbing matters which that debate brings to the surface.

2. Determinism and Scientific Explanation

Human actions make up a rather generous and unmysterious class of observable phenomena. It has often been urged that many such actions cannot be explained scientifically. For if they could be scientifically explained, they would be determined. But if they are determined, they would not be voluntary and, indeed, a great many human actions are voluntary. So here is one place where limits on scientific explanation must be drawn.

This argument has three premises. One asserts a connection between scientific explanation and determinism, another asserts a connection, or rather the lack thereof, between determinism and voluntary action, and the remaining premise asserts the (presumed) fact that there are voluntary human actions. Sometimes the argument is given an ethical twist by supplementing the premises with another premise to the effect that if human actions weren't voluntary, persons couldn't be held morally responsible for their actions. But we will be concerned mainly with the original and simpler argument.

The conclusion, indeed, seems to follow from the premises. So the question arises whether the argument is sound, whether, that is, all the premises are true and hence whether the argument *proves* its conclusion.

Voluntary actions are actions under the control of the person who performed them. A reflex action such as the startle response is not a voluntary action, but pressing the accelerator in your car is. Though not completely uncontroversial, the belief in the existence of voluntary human actions seems well founded. So the soundness of the argument that some human actions are immune to scientific explanation depends on the other two premises. We shall take them up in order.

The *first* premise states a relationship between scientific explanation and determinism, but what that relationship is is not very clear because determinism can be understood in a variety of ways. Two are particularly relevant to our discussion. The first, or "weak" version of determinism holds that an event or state-of-affairs is determined if it is "covered" by a law; the second or "strong" version of determinism holds that an event or state-of-affairs is determined if it is covered by a *causal* law. On either interpretation of determinism the first premise is false from the point of view of the pragmatic theory of explanation because that theory denies that scientific explanations *require* that the states-of-affairs being explained be covered by a law. Moreover, the premise is false from the point of view of the classical theory of scientific explanation, *if* determinism is taken in the strong sense, because that theory does not require that the covering laws be causal laws.

The question arises, therefore, whether covering law theories of scientific explanation can accommodate voluntary human actions. The answer to this question lies in the credibility of the second premise, the premise that states a relationship between determinism and voluntary action, because there is at least one sense of "determinism" in which the first premise is true in either the classical or the causal statistical theories of scientific explanation, both of which are covering law theories.

Let us, first, inquire further into the concept of voluntary action. What does it mean to say that an action is under one's control? It means that had the person performing the action so chosen, he *could have done otherwise*. So construed, the claim that determined actions are nonvoluntary, in the weak sense of "determinism", is very implausible. In weak determinism, the sense sponsored by the classical theory of scientific explanation, the covering laws can be of the noncausal statistical variety. Such laws, as pointed out in the chapter on explanation, can be ingredients in inductive explanations, the sort of explanation which does not rule out as *impossible* the nonoccurrence of the state-of-affairs being explained. Take as a sample state-of-affairs President Truman's decision to drop atomic bombs on Japan. There is no conflict in saying that there is an inductive explanation of his decision, in terms of noncausal statistical laws concerning what most people would do in a similar situation, and in maintaining that this decision was voluntary. For an inductive explanation of this kind is compatible with saying that had Truman so chosen, he *could* have done otherwise and refused to have the bombs dropped.

The question now arises whether, even in the strong sense, determinism is compatible with an action's being voluntary (and hence, again, whether the second premise of the argument that some human actions are scientifically explainable is false). The question is vital because the causal statistical theory requires that the explanans of a scientific explanations contain causal laws.

It is not at all clear that if the causal laws are statistical in character—such as the law about crime in the slums cited in Chapter II—that there is a conflict between an action's being determined and its being voluntary because it is not clear that its being so determined rules out the possibility that had the performer so chosen, the action could have turned out otherwise. Nor is it clear that there is a conflict *even if the causal law is nonstatistical* and thus asserts an invariable relationship between occurrence of the cause and occurrence of the effect. In this case there are two different sorts of argument to show that strong determinism and "freedom" of action are not in conflict. One derives from Aristotle; the other, from Immanuel Kant.

Aristotle argued that an action is voluntary, not when it is uncaused, but rather when the cause of the action "lies within." The appropriate causes are beliefs, deliberations, reasons, desires, and so on, and not, for example, muscle spasms. Thus if the *appropriate* causes are cited, the alleged conflict between determined and voluntary action evaporates.[1]

This way of reconciling determinism and voluntary action has much to recommend it. First, it connects, in a very intuitive way, voluntary action with a performer's desiring to do something. Second, and again intuitively, it rules out all those actions caused by "external" circumstances, hypnosis, coercion, and so on. Third, it makes sense of the fact that we have control over many of our actions.

Let us examine this last consideration in a little more detail. Too often those who insist on the voluntary character of some human actions suggest that such actions are completely spontaneous, uncaused, even irrational or random. But the notion of voluntary action seems to include the fact that such actions are under one's control. Indeed, the courts do not hold a person responsible for actions that are uncontestably irrational or random. The operative causes of actions under our control "lie within."

Vis-a-vis the causal statistical theory of scientific explanation, the theory which is now the focus of concern, the main difficulty with this ingenious account is that it is a controversial matter whether explanations in terms of inner causes qualify as scientific. The reason, as was detailed in the section on intentional explanations in Chapter II, is that the generalizations linking desires, beliefs, deliberations and so on with

[1]This is an oversimplified rendition of Aristotle's argument in the *Nichomachean Ethics.*

actions are arguably *not lawlike*, but the causal-statistical theory is a covering law "model" of scientific explanation. So, to put the point in a nutshell, whether the Aristotelian effort to reconcile voluntary action with determinism (in the strongest sense) succeeds depends, in large part, on whether intentional explanations—explanations in terms of "inner causes"—are genuinely scientific.

The other classic argument to reconcile voluntary action with strong determinism was proposed by Immanuel Kant.[2] For Kant (who, incidentally, conceived the principal task of philosophy to be determination of the limits of scientific explanation) the essential problem was to reconcile two fundamental principles, the principle that ought implies can—that is, that ascription of moral vocabulary to actions entails that the actions so characterized are voluntary (which Kant took to mean "uncaused")—with the principle that every event has a cause. Because Kant believed actions to be events, the two principles seem, on the face of it, to be incompatible.

Kant's solution, roughly stated, was to maintain that actions are describable in two different ways

> (1) as physical events to which the vocabulary of science is appropriate, and
> (2) as mental or intentional events to which the moral vocabulary is appropriate.

There are not two different sorts of events, only two different sorts of descriptions. The point is that whether a given action is voluntary is a function of the way it is described. But this does not conflict with the conviction that that action is also determined because the property of being determined applies to the action in question only when it is described in physical terms.

An illustration may help. One and the same event can be described as *a person's raising one's arm* or as *a person's arm going up*. The first description is intentional; the second is physical. Under the second description a causal statistical explanation in terms of muscle contraction, etc. is available, but under the first description it also makes good sense to say that the action described as the raising of a person's arm is under that person's control. Thus the conflict between an action's being

[2]The current interpretation of Kant's position relies heavily on Donald Davidson's essay, "Mental Events" in *Experience and Theory* (Ed. Lawrence Foster) Univ. of Massachusetts Press (1970). The relevant Kantian texts are *The Foundations of the Metaphysics of Morals* and the section on the third antimony in *The Critique of Pure Reason.*

determined and its being voluntary evaporates.

Kant's argument has its problems despite its compelling character. First, it supposes that one can give conditions for identifying events and say exactly what "appropriate description" means. Neither matter is easily solvable. Second, because, under the Kantian picture, science explains events only under "appropriate descriptions" it leaves something out—their mental or intentional aspects. So it isn't events or actions that resist scientific explanation but certain aspects of these actions, namely, their *mental* or *intentional* properties. It is these properties, and not certain actions, that defy scientific explanation via the causal statistical theory.

It is time to sum up the results of the discussion whether scientific explanation of voluntary action is possible. First, the view of scientific explanation one chooses greatly affects the course of the argument. Second, even in the most demanding of the three theories, the causal statistical theory, it is not a foregone conclusion that voluntary human actions are not scientifically explainable.

3. Reduction, Unification, and Scientific Explanation

Some critics of scientific explanation hold that certain phenomena are "emergent" with respect to others and so cannot be scientifically explained by the latter. For if they could be so scientifically explained, then the former could be "reduced" to the latter. In particular, biology could be reduced to physics, mind to matter, life to inanimate matter. But emergent phenomena are irreducible phenomena. So they cannot be scientifically explained.

Despite its long history, the argument is unclear. What, for example, does it really mean to say that a phenomenon cannot be reduced to other phenomena? Among emergentists reduction is usually understood along lines mentioned in the chapter on theories; rather than talk directly about reduced phenomena, they talk indirectly about them via theories about phenomena. A theory T_1 *reduces* to another theory T_2 if T_1 can be derived, possibly with the help of additional assumptions, from T_2. For example, the phenomena of heat reduce to mechanical phenomena because the thermodynamical theory that applies to the former can be derived from the statistical theory that applies to the latter via the assumption that temperature is mean kinetic energy. Another oft-cited example is the reduction of genetics to biochemistry via the Watson-

Crick assumption that the gene is the DNA molecule.[3]

The basic relation, then, between emergent and reduced phenomena is expressed in the *axiom*: if the theory about a phenomenon *P* is derivable from the theory about phenomena Q, then *P* is not emergent with respect to Q. For instance, the odor of ammonia gas emerges from the electrified combination of nitrogen and hydrogen (in appropriate quantities), because the theory about this odor is not derivable from the chemical theory *about* nitrogen and hydrogen. If it were so derivable, that is, if the odor of ammonia gas were so reducible, then that odor would not be emergent with respect to the electrified combination of nitrogen and hydrogen.[4]

The structure of the argument from emergents, apparently, is as follows:

> (1) Any phenomenon *P* scientifically explainable by phenomena *Q* is reducible to *Q*;
> (2) Any phenomenon *P* which is emergent with respect to phenomena *Q* is irreducible to *Q*;
> (3) So, any phenomenon *P* which is emergent with respect to *Q* is not scientifically explainable by *Q*.

This argument is valid. However, even granting its soundness for the moment, it establishes a far weaker conclusion than the emergentist's rhetoric demands. For it may well be the case that an emergent phenomenon not scientifically explainable by certain phenomena may nevertheless be scientifically explainable with respect to *other* phenomena—perhaps in virtue of being reducible to those other phenomena. An example of Ernest Nagel will corroborate the point.

> ...behavior consisting in variation in [a] clock's temperature or in changes in magnetic forces that may be generated by the relative motions of the parts of the clock—is not explained or predicted by mechanical theory. However, it appears that nothing but arbitrary custom stands in the way of calling these "nonmechanical" features of the clock's behavior "emergent properties" relative to mechanics. On the other hand, such nonmechanical features are certainly explicable with the help of theories of heat and magnetism, so that, relative to a wider class of theoretical

[3]There is an excellent account of their discovery in Watson's book, *The Double Helix*, Mentor (1968).

[4]The example is C. D. Broad's, one of the leading figures in the discussion. See his book, *The Mind and Its Place in Nature*, Routledge and Kegan Paul (1925).

assumptions, the clock may display no emergent traits.[5]

The upshot is that emergence is a *relative* notion; what is emergent with respect to one set of phenomena may be nonemergent with respect to other sets of phenomena and, thus, after all, scientifically explainable by the latter.

But is the argument from emergents, as outlined above, in fact sound? The answer to this question depends on the truth of premise (1); the truth of the second premise is guaranteed by the meaning of "reducible" and the axiom relating derivability with emergence.

We shall proceed by the method of example, and the example we shall fix on is the explanation of the occurrence of rain in Salzburg mentioned in the next to last section of the chapter on explanation. This example, it will be recalled, qualifies as a scientific explanation in all of the theories of scientific explanation discussed in this book.

The phenomenon to be explained is the occurrence of rain in Salzburg and the explaining phenomena are the existence of a low pressure area in the Salzburg region prior to the occurrence of rain, the temperature of the surrounding air, etcetera. It seems plausible to assume—and indeed in this example it seems indisputable—that the meterological theory *covering* the occurrence of rain is the same as the theory *about* low pressure areas, air temperature, etcetera. Now it is trivially true that a theory is *derivable from itself* and, therefore, given the emergentist's meaning of reduction, that the occurrence of rain in Salzburg reduces to the occurrence of a low pressure area in Salzburg prior to the occurrence of rain, the temperature of the surrounding air, etcetera.

To say that the occurrence of rain *reduces* to the existence of a prior low pressure area, the surrounding air temperature, etcetera, is, however, distinctively odd. If one looks at the typical cases of reduction in science—the reduction of thermodynamics to statistical mechanics, for example—it seems to be a necessary condition of reduction that the theory being derived (thermodynamics) not be the same as the theory from which it is derived (statistical mechanics). The reduction is effected by the *discovery* that the temperature of a *gas* can be identified with the mean kinetic energy of the *molecules making up the gas*; the theories, in other words, are not only different *vis a vis* their forms of expression but also *vis a vis* their subjective matters. This suggests that the account of reduction favored by emergentists must be amended to require that the theories T_1 and T_2 be distinct. But then the example explanation of rain

[5]Nagel, E. *The Structure of Science*, Hackett Publishing Co., p. 373.

in Salzburg does not imply reducibility—thus falsifying premise (1)—because the theories covering both explained phenomenon and explaining phenomena are the same. The dilemma provoked by the argument from emergents is clear: Either premise (1) is false or the emergentist's view of reduction is, at the very least, odd. This dilemma only reemphasizes the remarks at the beginning of the second paragraph of the section to the effect that despite its vintage, the argument from emergents is unclear.

Perhaps the concern of the emergentist can be better understood from a slightly different perspective. As noted in the introductory chapter, one of the leading themes of the classical position is the "unity of science." This phrase has several meanings. It can mean that science has a *methodological* unity. Many philosophers believe that if the word "science" has a meaning at all, it refers to a certain characteristic method. Sometimes this method is understood very generally, as in the injunctions "Respect the facts" or "Keep searching for the truth." Sometimes, as here, the method is understood in terms of modes of explanation and confirmation. Then the methodological unity of science means that a pattern of explanation, the classical theory, for example, is common to all the sciences.

The phrase "the unity of science" can also mean that science has a *structural* unity, but "structural unity" itself has at least three different senses. One is contained in the "layer cake" view of theories. The "layer cake" view, recall, asserts that knowledge can be arranged in a deductive structure, the most general and all-embracing theories at the top and the most directly empirical generalizations at the bottom. Another sense of "structural unity" is that each branch of science is associated with a particular kind of object, the branches arranged in a hierarchical order in terms of the complexity of these objects. At the bottom of the hierarchy stand the objects of physics, currently mesons and others, of which the objects studied by the other branches are ultimately composed.[6] Thus, psychology studies human behavior. But men are composed of cells. Since biology studies cell behavior, it is a more basic science than psychology, in the sense that the explanations it provides are more profound. But cells are composed of submicroscopic particles. Hence physics, which studies the behavior and simple properties of these particles, is ultimately more explanatory than biology. This sense of

[6]See Paul Oppenheim and Hilary Putnam, "Unity of Science as a Working Hypothesis," *Minnesota Studies in the Philosophy of Science*, Vol. II, eds., Feigl, Scriven, and Maxwell, University of Minnesota Press (1955).

"structural unity" dates from the Greek philosophers; to explain all natural phenomena in terms of the properties of their least parts has been a goal of science for a very long time.

A third sense of "structural unity" concerns the apparent fact that the passage of time has brought with it physical explanations of ever more phenomena. It was thought throughout the 19th century, for example, even by many competent biologists, that organic phenomena could never be explained in entirely physical terms; at the very least, some sort of entelechy or "vital spirit" had to be postulated to understand the behavior of living organisms. But the 20th century, and the advent of molecular biology, has completely undermined this belief. Organic phenomena too can be provided with physical explanations.

Because the first sort of structural unity is very closely related to derivational reduction, we call it *logical unity*. The second sort of structural unity is called, in contrast, *part/whole unity*. The third sort of structural unity we call *progressive unity*.

The point of this lengthy discussion of structural unity is simply that the emergentist denies, in *all* its forms, the structural unity of science. There are phenomena which cannot be fitted into the general derivational pattern, or which cannot profitably be decomposed into their subatomic parts, or which resist the best efforts of scientists to provide a purely physical explanation for them. It is in these ways that the emergentist sets limits to science. It is from this perspective that we can understand best the emergentist's claim that certain phenomena are "irreducible."

Two facts seem to stand out. One is that the classical account of reduction as a deductive relationship between theories, and the associated view that science has a logical unity, give rise to problems concerning the nature of the connection between reduced and reducing theories. If the one theory is to be derived from the other, there must be a way of connecting terms in the two theories but the emergentist insists that there is no *satisfactory* way of doing this (i.e., the condition of connectability cannot be satisfied). The other fact is that there has been some sort of unification in the history of science, at least of the kind we have called progressive unification. This is part of what is usually meant by talk about the *progress* that science has made over time. Sharp boundaries between the chemical and the mechanical, between the organic and the inorganic, even between the mental and the physical, have gradually given way. The emergentist who denies such unification is simply wrong and thus there must be a flaw somewhere in his argument. Nancy Maull has suggested a promising new view of reduction which takes account of these two facts and undermines both the pretensions of the emergentist and the classical picture of reduction.[7]

Maull begins with a critique of the classical account of reduction.[8] In the first place, it blurs an important distinction between *theories* and *branches* of science. The classical account of reduction suggests that to reduce one branch of science, e.g., biology, to another, e.g., physics, is simply to reduce one theory to another. But theories and branches of science cannot be equated; in fact, some branches of science, notably biology, lack a comprehensive theory. In the second place, the classical account is not a very accurate, still less a very useful, description of what actually takes place when there is theoretical unification. In the third place, the classical account implies that progress in the history of science is marked by the development of more and more *general* theories; the ideal goal is a theory which holds all natural phenomena in its deductive embrace. But, again to take biology, for example, the trend has not invariably been towards more general theories. We now have a unified treatment of biological, chemical, and physical phenomena, but it has nothing to do with reduction to more general theories.

Nor does Maull think that unification in science obeys the part/whole model.[9] On the part/whole model, branches of science are characterized in terms of the objects which they study. Unification takes place when objects on one "level" are decomposed in terms of objects on the next level "below" them. Putnam and Oppenheim, in "Unity of Science as a Working Hypothesis," identify six such "levels": social groups, multicellular living things, cells, molecules, atoms, elementary particles. The difficulty, according to Maull, is that branches of science cannot be assigned without further question to a unique "level." The apparent place to assign genetics, for example, is the molecular level since genes are

[7]See her "Unifying Science Without Reduction," *Studies in History and Philosophy of Science*, 8 (1977), pp. 143-62.

[8]Other difficulties with the deductive account of theory reduction can be found in papers by Lawrence Sklar, "Types of Inter-Theoretic Reduction," *British Journal for the Philosophy of Science*, 18, (1967), pp. 109-24, K. F. Schaffner, "Approaches to Reduction," *Philosophy of Science*, 34 (1967), pp. 137-47 and, from the "subjectivist" viewpoint, Paul Feyerabend, "Explanation, Reduction, and Empiricism," in Feigl and Maxwell, eds., *Minnesota Studies in the Philosophy of Science*, University of Minnesota Press (1962), Vol. III. A good statement of Feyerabend's point of view on reduction and its relation to his theory of theoretical change discussed earlier can be found in Chapter 11 of I. Hacking's *Why Does Language Matter to Philosophy?*, Cambridge University Press (1972).

[9]For a discussion of some of the leading features of the part/whole model and its relation to the classical account, see Gordon G. Brittan, Jr., "Explanation and Reduction," *Journal of Philosophy*, LXVII (1970), pp. 446-57.

molecules. But genetics is not a molecular science, a fact which suggests assignment to still another "level." The source of the difficulty is the fact that what are apparently the "same" parts and wholes are investigated at different levels and in different terms, always with respect to a field-defining problem. The genes studied in genetics, for example, are identified with the DNA sequences studied in biochemistry, each described from a characteristic point of view.[10] Maull's positive account of unification is rooted in a study of these problems, for it is problems which define a branch of science (or "field"),[11] along with characteristic methods and techniques. Further, the important interactions in science leading to unification have not been between theories but between fields. Now the striking fact about problems is that they shift. I.e., problems arising within one field often become the problems of another field, in part because they cannot be solved using the concepts and techniques of the field in which they arise. Thus problems which arose within genetics concerning the physical nature of the gene were solvable only by way of the concepts and techniques of biochemistry, while the discovery that the gene is the DNA molecule posed problems solvable in turn by physical chemistry. It is the solutions to such problems, and the sharing of vocabulary they presuppose, which provides connections between theories or, more broadly, between fields. But these solutions are provided by theories which themselves do the connecting. Thus, instead of one theory being "reduced" to another, two theories or fields are unified by way of the development of a third, "interlevel," theory or field.

In the final analysis, unification in the history of science is of the *progressive* sort. New theories arise in the course of time to unify hitherto disparate domains. The unification of genetics, biochemistry, and physical chemistry around which the discussion has been focused may be taken as typical. The emergentist, though right about the defects of a reductive account of unification, misses the larger points (a) that unification, independent of reduction, is still possible, and (b) that such unification is characteristic of scientific progress. It doesn't follow that unification is inevitable or that all natural phenomena will eventually be fitted into the scientific picture. What follows is only that the failure of the reductive account doesn't bar the door to such developments.

[10]"Unifying Science Without Reduction," p. 155.

[11]Thus genetics is the field whose characteristic problem is gene difference, psychology is the field whose characteristic problem is behavioral difference, etc. For more on the general concept of "fields," see Lindley Darden and Nancy Maull, "Interfield Theories," *Philosophy of Science*, 44 (1977), pp. 43-64.

Index

VILLAGE CENTENARY

MISS READ

Illustrated by J. S. Goodall

Houghton Mifflin Company Boston

1981

First American Edition 1981

Library of Congress Cataloging in Publication Data

Read, Miss.
Village centenary.

I. Title.
PR6069.A42V5 1981 823'.914 81-6300
ISBN 0-395-31262-0 AACR2

Printed in the United States of America

V 10 9 8 7 6 5 4 3 2 1